# Endangered Marine Vertebrates - Recent Advances for Conservation

*Edited by Mohamed Nejmeddine Bradai, Imed Jribi, Bechir Saidi and Samira Enajjar*

Published in London, United Kingdom

Endangered Marine Vertebrates - Recent Advances for Conservation
http://dx.doi.org/10.5772/intechopen.1003504
Edited by Mohamed Nejmeddine Bradai, Imed Jribi, Bechir Saidi and Samira Enajjar

Contributors
Ahmed A. Mustafa, Azemat Dadai Ghandi, Bechir Saidi, Cristina Pedà, Danilo Malara, Foong Swee Yeok, Imed Jribi, Ismail M. Abdelazim, Mahmoud M.S. Farrag, Manel Khammassi, Mohamed Nejmeddine Bradai, Mohammad Sayad Bourani, Samira Enajjar, Siamak Bagheri, Yassein A. Osman



First published in London, United Kingdom, 2025 by IntechOpen
IntechOpen is the global imprint of INTECHOPEN LIMITED, registered in England and Wales, registration number: 11086078, 167-169 Great Portland Street, London, W1W 5PF, United Kingdom

For EU product safety concerns: IN TECH d.o.o., Prolaz Marije Krucifikse Kozulić 3, 51000 Rijeka, Croatia, info@intechopen.com or visit our website at intechopen.com.

British Library Cataloguing-in-Publication Data
A catalogue record for this book is available from the British Library

Endangered Marine Vertebrates - Recent Advances for Conservation
Edited by Mohamed Nejmeddine Bradai, Imed Jribi, Bechir Saidi and Samira Enajjar
p. cm.
Print ISBN 978-0-85014-933-3
Online ISBN 978-0-85014-932-6
eBook (PDF) ISBN 978-0-85014-934-0

If disposing of this product, please recycle the paper responsibly.

# Meet the editors

Mohamed Nejmeddine Bradai is a Professor of Higher Education, a former researcher at the National Institute of Marine Science and Technologies, and the head of the Marine Biodiversity Laboratory. Currently, in retirement, he continues research mainly in the frame of an environmental NGO, "Association de Protection de la Biodiversité dans le Golfe de Gabès". He has an engineering degree in marine living resource exploitation. He has prepared three post-graduate degrees (Master's, Doctorate (Ph.D.) and State Doctorate. He is working mainly on the biodiversity of marine vertebrates (Sea turtles, sharks and cetaceans). Pioneer researcher in the conservation biology of sea turtles and sharks in Tunisia and the Mediterranean Sea. Dr. Bradai is a member of the Shark Specialist Group and Sea Turtles Specialist Group of the IUCN and a former coordinator of the Sub Committee "Marine Environment and Ecosystems» of the Scientific Advisory Committee (SAC) of the GFCM-FAO.

Dr. Imed Jribi is an Associate Professor in Biological Sciences at the University of Sfax, Tunisia, specializing in marine ecology and conservation. With more than 25 years of experience, his research focuses on the biology, ecology, and conservation of mainly sea turtles but also on cetaceans and fish. He has led numerous national and international projects, contributed to over 60 peer-reviewed publications, and actively participated in marine biodiversity conservation initiatives. Dr. Jribi serves on multiple scientific committees, including the IUCN Marine Turtle Specialist Group, and has extensive field experience in monitoring nesting beaches and studying marine species interactions. His expertise extends to marine policy development and training programs across the Mediterranean region.

Bechir Saidi is a marine biologist who obtained his Ph.D. from the University of Sfax (Tunisia) in the taxonomy, biology, and fishery studies of sharks. Since 2002, he has been a member of the Marine Biodiversity Laboratory at the National Institute of Marine Sciences and Technologies (INSTM), where he has been conducting research on the elasmobranchs to assess the impact of anthropic-related activities on their conservation. He has been involved in several national and international projects dealing with elasmobranch biology, taxonomy, and fisheries interactions with the aim of producing comprehensive data to promote conservation action. He is a member of the Shark Specialist Group of the IUCN.

Samira Enajjar is a marine biologist affiliated with the Marine Biodiversity Laboratory of the National Institute of Marine Sciences and Technologies (INSTM), Tunisia. She has been researching eco-biology, biodiversity, fishery, and taxonomy of elasmobranchs, mainly of batoids in the Gulf of Gabès (Southern Tunisia, central Mediterranean Sea). Dr. Enajjar contributed to the elaboration of vulnerable marine vertebrate training guides. She is also a member of the Med Bycatch Project as a supervisor and expert on elasmobranchs. She is regularly engaged in training courses on the taxonomy of rays in many national and regional projects. She is involved in developing the Mediterranean Angel Sharks regional and sub-regional action plans.

# Contents

# Preface

Although they have a widespread global distribution, most known vertebrate species are threatened. These taxa are ecosystem engineers that contribute to preserving marine and coastal habitats. For example, leatherback turtles are one of the main predators of jellyfish, reducing the impact of predation by these cnidarians on copepods and fish larvae. Similarly, hawksbill and green turtles feed on sea sponges and seagrasses, keeping their development under control in a way that maintains the health of coastal habitats such as coral reefs and enhances the reproduction of marine-associated fish and shellfish species.

Cetaceans contribute to the biological and chemical cycles of the ocean by dispersing nutrients through their excrement, playing a crucial role in nutrient circulation.

Sharks, as apex predators, have few natural enemies and regulate marine ecosystems by feeding on weaker, older, and sick prey. This process helps maintain the overall health of prey populations.

Sharks also play a vital role in preserving the evolutionary 'good health' of the species they prey on. No other predator can fully replace sharks in marine food chains. Their decline leads to a decrease in small fish populations, affecting coral reef health and reducing oxygen production in the ocean.

Despite their essential ecosystem services, threatened marine vertebrates are exposed to various human-induced pressures. Their incidental capture (Bycatch) in many fishing gear is considered the major threat. In addition to bycatch and the impact of fishing activities, other threats include marine litter pollution, mainly plastic, loss of essential habitats and climate change.

Marine litter, mostly consisting of plastics, is ubiquitous in the environment while several million tons of plastic continue to enter the ocean every year, and current densities could triple by 2040 (UNEP, 2021). The damage to the marine environment is alarming and has possible cascading impacts on natural resources.

Ocean predators keep marine life populations in check to prevent an upset of the ecological balance. Overfishing poses serious threats as the loss of predators, like tuna and sharks, allows populations of prey species to expand. This then can lead to a destabilized food web and marine environment.

Moreover, the interplay of some traits, such as ontogenetic habitat shifts, long-distance migrations, and temperature-dependent sex determination (TSD) in sea turtles, make good models for studying and assessing the impact of climate change, a threat that is still poorly understood despite the increased attention of policymakers, scientists, and the general public.

Maintaining and restoring populations of large marine vertebrates appears to be a more credible nature-based solution than large-scale artificial fertilization to promote carbon capture by the ocean.

This book explores various threats to iconic vertebrates and discusses recent advances and conservation strategies to mitigate their impact.

I extend my gratitude to all the chapter authors for their valuable contributions. I am also deeply thankful to my three co-editors, each of whom is an expert in at least one of the taxa covered in this book—fishes, sea turtles, and marine mammals.

Finally, I would like to acknowledge the team at IntechOpen for their support throughout the editorial process.

**Mohamed Nejmeddine Bradai**
ASCOB-Syrtis,
Faculty of Sciences of Sfax,
Sfax, Tunisia

**Imed Jribi**
Faculty of Sciences of Sfax,
University of Sfax,
Sfax, Tunisia

**Bechir Saidi**
Faculty of Science and Technology of Sidi Bouzid,
University Campus Agricultural City,
Sidi Bouzid, Tunisia

**Samira Enajjar**
National Institute of Sciences and Technologies of the Sea (Sfax Center),
Sfax, Tunisia

Section 1

# Knowledge of Ecology and Life History for Management and Conservation Options

Chapter 1

# Overview of the Angel Sharks in the Mediterranean Sea

*Samira Enajjar, Bechir Saidi and Mohamed Nejmeddine Bradai*

## Abstract

Three species of angel shark are present in the Mediterranean Sea with overlapping ranges: The Sawback angelshark *Squatina aculeata*, the Smoothback angelshark *Squatina oculata* and the angelshark *Squatina squatina*. All three are classified as critically endangered according to the IUCN due to past population reductions and are listed under the binding recommendation GFCM/42/2018/2 (amending GFCM/36/2012/3). This recommendation prohibits the retention and sale of the three species. A recent survey confirms the dramatic decline of their populations in all of the Mediterranean Sea. The compilation of bibliographic and unpublished data show that only 280 angelsharks were sighted (all species confused) during the last 30 years. The eastern (GSA 22) and the central Mediterranean Sea (GSA 12, 13, 14 and 15) represent an important area for Squatinidae species. All life stages were reported; however, maturing specimens dominate the sighting individuals. Bottom trawl and trammel nets were the gear most affected by the Squatinidae population due to their benthic and shallow habitat range. Considering the status of the species and their life history parameters, Squatinidae are highly vulnerable to fisheries. Therefore, immediate measures should be taken to conserve these species. These measures must be prioritized on: fisheries, habitats and legislation.

**Keywords:** angelsharks status, fishery, critical habitats, conservation, Mediterranean sea

## 1. Introduction

The Mediterranean Sea is said to have the highest percentage of endangered sharks in the world. Despite their relative diversity in the area (88 species) and the almost absence of target fisheries, more than 53% of species are endangered (critically endangered, endangered and vulnerable), according to the IUCN [1]. All species are fished mainly as bycatch, highlighting the negative effect of irresponsible fisheries on the populations of this taxa. It has been confirmed that sharks in the area have declined by more than 97% in number and "catch weight" over the last 200 years [2]. Likewise, the main elasmobranch-fishing countries in the area register a dramatic decline in landing [3, 4]. Consequently, the Mediterranean Sea is designated as a key hotspot of extinction risk for chondrichthyan species [1].

In this chapter, we analyze bibliographic and unpublished data (present study) of Squatinidae in the Mediterranean Sea, an example of a critically endangered species, in order to investigate their status and to propose a conservation strategy. In fact, angel sharks are considered as the third most threatened family of elasmobranchs in the world. The life history and ecology of most angel sharks remain poorly known and their life characteristics are fragile: slow growth, late maturity and low reproductive rate [5]. Their demersal nature as well as the overexploitation and the illegal fishery, make angelsharks population threatened.

## 2. Study area

The Mediterranean Sea is a semi-enclosed sea located between three continents: Africa, Asia and Europe. It connects to the Atlantic Ocean by the narrow and shallow channel of the Strait of Gibraltar, which is roughly 13 km wide at its narrowest point. To the northeast, the Mediterranean Sea is linked to the Black Sea through the Dardanelles, the Sea of Marmara and the Strait of the Bosporus. To the southeast, it is connected to the Red Sea by the Suez Canal [6].

Mediterranean hydrodynamics are driven by three layers of water masses: a surface, an intermediate and a deep layer that sinks to the bottom; a separate bottom layer is absent.

A submarine ridge between the island of Sicily and the African coast, with a sill depth of about 365 meters, divides the Mediterranean Sea into western and eastern sub-basins [7].

The salinity of the Mediterranean is uniformly high throughout the basin. Surface waters average is about 38‰ except in the extreme western parts; the salinity can approach to 40‰ in the eastern Mediterranean during the summer. Deepwater salinity is slightly less than 38.4‰ [8].

The highest temperature in the Mediterranean is recorded in the Gulf of Sidra (Libya), where the mean temperature in August is about 31°C. The lowest surface temperatures are found in the extreme North of the Adriatic, where the mean temperature in February falls to 5°C (Gulf of Trieste) [8].

The Mediterranean suffers from intensive exploitation of fishery resources. The overexploitation is strengthened by the use of trawl nets with very small mesh sizes that retain the smallest individuals and cause the bycatch of vulnerable species (sea turtles, cetaceans and elasmobranchs) [9].

## 3. Data collection

In order to access the Squatinidae status in the Mediterranean Sea, a database containing all the confirmed species records in the basin during the last 30 years was compiled using diverse sources: (1) bibliographic citations, (2) unpublished records collected by authors from direct surveys of landings or reported from a project developed in the area, (3) citizen sciences through social media including photographs of captured angel sharks. The validity of the record has been confirmed by means of direct contact with the fisherman reported in the source. Any doubtful records or lacking relevant details were not considered in this work.

For each record, date, location with associated physiographical covariates (distance from shore and depth), fishing gear, total length TL (cm), sex, uteri content, embryo numbers, sizes, etc., were noted when available. In most cases, a photograph of the specimen capture is available.

The recent increase in research efforts in the Mediterranean Sea revealed numerous new records of all three angelshark species. A total of 280 angelsharks sightings (all species confused) were recorded in the Mediterranean water from historical and contemporary observation (**Tables 1–3**). The number of Squatinidae examined exceed 318 specimens. Nevertheless, more than 71% of signaled specimens were registered after 2010, which is the period when specific attention was concentrated on angelsharks and many projects were conducted in the area.

| Year | Month | GFCM sub-area | Sex | Total length (cm) | Gear | Depth | References |
|---|---|---|---|---|---|---|---|
| 1980 | June | 19 | | 110 | | | [10] |
| 1993 | August | 16 | F | 29,2 | Bottom trawl | 250 | [10] |
| 1997 | February | 14 | M | 80 | Bottom trawl | | [11] |
| 1997 | February | 14 | M | 45.4 | Bottom trawl | | [11] |
| 1997 | July | 22 | M | 30 | Bottom trawl | | [12] |
| 1999 | September | 22 | F | 95 | Bottom trawl | | [12] |
| 2002 | Spring | 28 | | | | | [13] |
| 2004 | July | 13 | M | 79,5 | Gillnets | | Present work |
| 2004 | September | 12 | F | 98 | Longline | | Present work |
| 2004 | September | 12 | F | 102 | Bottom trawl | | Present work |
| 2004 | June | 13 | F | 78 | Bottom trawl | | Present work |
| 2004 | January | 14 | F | 69,5 | | | Present work |
| 2004 | January | 14 | F | 68 | | | Present work |
| 2004 | January | 14 | M | 77 | | | Present work |
| 2004 | December | 22 | F | 79,5 | Bottom trawl | 60–80 | [14] |
| 2005 | September | 15 | | 96 | | | [10] |
| 2005 | September | 15 | F | 156 | | | [10] |
| 2006 | February | 14 | F | 98 | | | Present work |
| 2006 | June | 15 | | | Bottom trawl | 126 | [10] |
| 2007 | September | 19 | 3 M/ 1 F | 29,1/44,2/ 43,25/56,4 | Bottom trawl | 50 | [15] |
| 2007 | November | 12 | F | >80 | Bottom trawl | 150 | [10] |
| 2007 | November | 12 | M | >80 | Bottom trawl | 150 | [10] |
| 2007 | November | 12 | M | >80 | Bottom trawl | 150 | [10] |
| 2007 | November | 12 | M | >80 | Bottom trawl | 150 | [10] |

| Year | Month | GFCM sub-area | Sex | Total length (cm) | Gear | Depth | References |
|---|---|---|---|---|---|---|---|
| 2008 | June | 15 | F | 42 | Bottom trawl | 145 | [10] |
| 2008 | June | 15 | F | 29,5 | Bottom trawl | 173 | [10] |
| 2008 | June | 15 | M | 32 | Bottom trawl | 173 | [10] |
| 2008 | June | 15 | M | 37,5 | Bottom trawl | 132 | [10] |
| 2008 | June | 15 | F | 94 | Bottom trawl | 80 | [10] |
| 2009 | Jan- Juin | 14 | F | 80 | Bottom trawl | | [16] |
| 2009 | Jan- Juin | 14 | F | 118 | Bottom trawl | | [16] |
| 2009 | August | 24 | | 80 | Bottom trawl | 100 | [17] |
| 2010 | September | 24 | | 50 | Bottom trawl | 100 | [17] |
| 2010 | September | 24 | F | 88 | Bottom trawl | 50 | [17] |
| 2010 | September | 24 | | 59 | Bottom trawl | 50 | [17] |
| 2010 | Winter | 24 | | 52 | Bottom trawl | 100 | [17] |
| 2010 | Winter | 24 | | 67 | Bottom trawl | 50 | [17] |
| 2010 | Winter | 24 | | 69 | Bottom trawl | 50 | [17] |
| 2010 | Winter | 24 | | 66 | Bottom trawl | 50 | [17] |
| 2010 | Winter | 24 | | 88 | Bottom trawl | 50 | [17] |
| 2010 | Winter | 24 | | 24 | Bottom trawl | 50 | [17] |
| 2013 | June | 15 | M | 74 | Bottom trawl | 101 | [10] |
| 2014 | April | 14 | F | <80 | Bottom trawl | 26 | [10] |
| 2014 | April | 14 | M | >100 | Bottom trawl | 47 | [10] |
| 2015 | July | 23 | | | Nets | 20 | [10] |
| 2016 | February | 16 | F | 70,5 | Bottom trawl | 120 | [10] |
| 2016 | June | 15 | M | 77,5 | Trammel net | | [10] |
| 2016 | June | 15 | M | 77,5 | Trammel net | | [10] |
| 2016 | June | 15 | F | 97 | Trammel net | | [10] |
| 2016 | June | 15 | | 71 | Trammel net | | [10] |
| 2016 | August | 16 | M | 110 | Trammel net | | [10] |
| 2016 | August | 15 | F | 100 | Underwater observation | 35 | [10] |
| 2016 | November | 24 | F | 72,6 | Bottom trawl | 65 | [18] |
| 2016 | June | 14 | 2 F and 1 M | 60 | Gillnets | | Present work |
| 2017 | March | 16 | M | 73,4 | Bottom trawl | 180 | [10] |
| 2017 | June | 16 | | 98 | Trammel net | 50 | [10] |
| 2017 | June | 15 | F | 31 | Bottom trawl | | [10] |
| 2017 | June | 15 | F | 29,5 | Bottom trawl | | [10] |
| 2017 | June | 15 | F | 29,5 | Bottom trawl | | [10] |
| 2017 | August | 16 | F | 90,3 | Bottom trawl | 50 | [10] |
| 2017 | December | 16 | | | Trammel net | 55 | [19] |
| 2018 | December | 12 | F | 130 | Gillnets | | [20] |

| Year | Month | GFCM sub-area | Sex | Total length (cm) | Gear | Depth | References |
|---|---|---|---|---|---|---|---|
| 2018 | March | 22 | F | 87,5 | Bottom trawl | 110 | [21] |
| 2018 | May | 15 | F | 94,5 | Bottom trawl | | [10] |
| 2019 | March | 15 | M | 122 | Bottom trawl | | [10] |
| 2019 | March | 15 | M | 75 | Bottom trawl | | [10] |
| 2019 | octobre | 15 | F | 29 | Bottom trawl | | [10] |
| 2020 | | 22 | | | | | [5] |
| 2020 | February | 22 | | | | | [22] |
| 2020 | April | 22 | | 79,8 | | | [22] |
| 2020 | octobre | 22 | F | 80 | | | [22] |
| 2020 | | | | | | | [23] |
| 2021 | March | 26 | M | 70 | Bottom trawl | 55–70 | [10] |
| 2021 | March | 26 | F | 49,2 | Bottom trawl | 55–70 | [10] |
| 2022 | May | 13 | F | 100 | | | Present work |
| 2022 | May | 13 | F | 110 | | | Present work |
| 2022 | February | 12 | | 135 | Bottom trawl | 100–150 | [24] |
| 2022 | February | 12 | | 140 | Bottom trawl | 100–150 | [24] |
| 2022 | February | 12 | M | 90,5 | Bottom trawl | 100–150 | [24] |
| 2022 | November | 22 | | 100 | Trammel nets | | [22] |
| 2022 | November | 22 | | | Trammel nets | | [22] |
| 2023 | January | 22 | | 60–80 | Trammel nets | | [22] |
| 2023 | January | 22 | F and M | | Trammel nets | 120–135 | [22] |
| 2023 | January | 22 | | 72 | Trammel nets | 60 | [22] |
| 2023 | January | 22 | F | 110 | Trammel nets | | [22] |
| 2023 | January | 22 | F | | Trammel nets | 188–194 | [22] |
| 2023 | January | 22 | | 65 | Trammel nets | 190 | [22] |
| 2023 | January | 22 | M | 55 | Trammel nets | | [22] |
| 2023 | January | 22 | F - M | 50–65 | Trammel nets | 200 | [22] |
| 2023 | March | 22 | F | | Trammel nets | | [22] |
| 2023 | February | 22 | | | Trammel nets | | [22] |
| 2023 | March | 22 | F - M | 67, 83 | Trammel nets | 91 | [22] |
| 2023 | March | 22 | F | | Trammel nets | | [22] |
| 2023 | March | 22 | F | 20–23,69 | Trammel nets | 160 | [22] |
| 2023 | May | 24 | F | 150 | Trammel nets | | [22] |
| 2023 | May | 22 | | 38 | Trammel nets | | [22] |
| 2023 | May | 22 | M | 57 | Trammel nets | 17 | [22] |
| 2023 | May | 24 | F | 77 | Trammel nets | 111 | [22] |
| 2023 | July | 22 | F | 50–60 | Trammel nets | 39 | [22] |
| 2023 | June | 22 | | 38–150 | Trammel nets | | [22] |

| Year | Month | GFCM sub-area | Sex | Total length (cm) | Gear | Depth | References |
|---|---|---|---|---|---|---|---|
| 2023 | August | 22 | F | 48 | Trammel nets | 183 | [22] |
| 2023 | Septembre | 24 | F | 61 | Trammel nets | 93 | [22] |
| 2023 | Septembre | 24 | M | 81 | Trammel nets | 89 | [22] |
| 2023 | Septembre | 24 | F | 78 | Trammel nets | 108 | [22] |
| 2023 | September | 24 | F | 82 | Trammel nets | 100 | [22] |
| 2023 | September | 24 | F | 69 | Trammel nets | 108 | [22] |
| 2023 | September | 22 | F | 39,5 | Trammel nets | 110 | [22] |
| 2023 | October | 22 | M | 55 | Trammel nets | 110 | [22] |
| 2023 | October | 24 | F | 55 | Trammel nets | 130 | [22] |
| 2023 | November | 22 | F | 50,5 | Trammel nets | 167 | [22] |
| 2023 | December | 22 | M | 60 | Trammel nets | 160 | [22] |
| 2023 | December | 22 | F | 75 | Trammel nets | 185 | [22] |
| 2023 | December | 22 | M | 55 | Trammel nets | 169 | [22] |
| 2023 | December | 22 | M | 60 | Trammel nets | 239 | [22] |
| 2023 | December | 22 | M | 62 | Trammel nets | 210 | [22] |
| 2023 | December | 22 | M | 62 | Trammel nets | 204 | [22] |
| 2023 | December | 22 | M | 70 | Trammel nets | 210 | [22] |
| 2023 | December | 22 | M | 67 | Trammel nets | 200 | [22] |
| 2023 | December | 22 | F | 42,5 | Trammel nets | 150 | [22] |
| 2023 | December | 22 | M | 60 | Trammel nets | 162 | [22] |
| 2023 | December | 22 | F | 62 | Trammel nets | 171 | [22] |
| 2023 | November | 13 | F | 100 | Bottom trawl | | Present work |
| 2023 | November | 12 | | 100 | | | Present work |

**Table 1.**
*The historical and recent records of the* Squatina oculata *off the Mediterranean Sea.*

| Year | Month | GFCM sub-area | Sex | Total length (cm) | Gears | Depth | References |
|---|---|---|---|---|---|---|---|
| 1995 | November | 28 | F | | | | [25] |
| 1996 | October | 22 | M | 120 | Longline | | [12] |
| 1996 | October | 22 | M | 152 | Longline | | [12] |
| 1996 | October | 22 | M | 140 | Longline | | [12] |
| 1997 | July | 22 | M | 75 | Gillnet | | [12] |
| 1997 | July | 22 | M | 83 | Gillnet | | [12] |
| 2001 | January | 14 | F | 130 | | | [11] |
| 2002 | June | 14 | F | 169 | | | [11] |
| 2002 | April | 14 | F | 74 | | | Present work |
| 2003 | June | 14 | F | 169 | | | Present work |

| Year | Month | GFCM sub-area | Sex | Total length (cm) | Gears | Depth | References |
|---|---|---|---|---|---|---|---|
| 2003 | March | 14 | F | 159 | | | Present work |
| 2003 | March | 14 | M | 84 | | | Present work |
| 2003 | March | 14 | F | 102 | | | Present work |
| 2003 | June | 14 | F | 70,5 | | | Present work |
| 2004 | January | 14 | F | 128 | | | Present work |
| 2004 | February | 14 | F | 84 | | | Present work |
| 2004 | April | 14 | F | 154,5 | | | Present work |
| 2004 | June | 13 | F | 110 | | | Present work |
| 2004 | June | 13 | F | 75 | | | Present work |
| 2004 | June | 13 | F | 112 | | | Present work |
| 2004 | June | 13 | F | 103 | | | Present work |
| 2005 | March | 14 | M | 98 | | | Present work |
| 2004–2005 | | 22 | | 26,5 | Trammel net | 30 | [26] |
| 2005–2008 | | 22 | F | 23,2 | Bottom trawl | | [27] |
| 2008 | April | 17 | | 33,5 | | | [28] |
| 2009 | January-juin | 14 | F | 67 | Bottom trawl | | [16] |
| 2009 | January–June | 14 | F | 94 | Bottom trawl | | [16] |
| 2009 | March | 22 | F | 160 | Gillnet | | [29] |
| 2013 | October | 22 | | 100 | Harpoon | 1,5 | [29] |
| 2014 | January | 22 | F | 174 | trammel net | 50 | [29] |
| 2014 | January | 28 | F | 174 | Trammel net | 50 | [25] |
| 2015 | February | 22 | F | 156 | trammel net | 20 | [30] |
| 2015 | February | 22 | F | 156 | trammel net | 20 | [29] |
| 2015 | March | 22 | F | 160 | Gillnet | | [29] |
| 2015 | December | 22 | F | 150 | Diving observation | 30 | [29] |
| 2016 | December | 17 | | 80–90 | Bottom trawl | | [28] |
| 2017 | November | 17 | | 60 | Bottom trawl | | [28] |
| 2017 | November | 17 | | 14 | Bottom trawl | | [28] |
| 2017 | June | 8 | | 50 | Trammel net | <5 | [31] |
| 2017 | June | 8 | | 40 | Trammel net | <5 | [31] |
| 2017 | October | 8 | | 120 | Trammel net | <5 | [31] |
| 2017 | July | 22 | | | Stranded | | [19] |
| 2017 | December | 21 | | 100 | Recreational fishing | | [19] |
| 2018 | April | 8 | | 80 | Trammel net | <5 | [31] |
| 2018 | April | 8 | | 80 | Trammel net | <5 | [31] |
| 2018 | May | 8 | | 60 | Trammel net | <5 | [31] |
| 2018 | May | 8 | | | Trammel net | <5 | [31] |
| 2018 | February | 21 | | 100 | Trammel net | 50 | [19] |

| Year | Month | GFCM sub-area | Sex | Total length (cm) | Gears | Depth | References |
|---|---|---|---|---|---|---|---|
| 2018 | February | 21 | | 110 | Trammel net | | [19] |
| 2018 | February | 21 | | 80 | Recreational fishing | | [19] |
| 2018 | March | 22 | | 150 | Bottom trawl | 236 | [19] |
| 2018 | March | 21 | | 100 | Recreational fishing | | [19] |
| 2018 | March | 22 | M | 114 | Longline for sea breams | 40 | [29] |
| 2018 | April | 22 | M | 90 | Gillnet | 30 | [29] |
| 2018 | February | 28 | F | 140 | Gillnet | 20 | [29] |
| 2018 | March | 22 | M | 91,5 | Bottom trawl | 110 | [21] |
| 2020 | | 22 | | | | | [19] |
| 2020 | June | 13 | F | 90 | Bottom trawl | | [32] |
| 2020 | February | 12 | F | | | | Present work |
| 2020 | | 21 | | | | | [23] |
| 2021 | April–June | 12 | 7 M | 80–140 | Bottom trawl | | [33] |
| 2021 | April–June | 12 | 9 F | 80–140 | Bottom trawl | | [33] |
| 2022 | July | 22 | | 30–40 | | | [22] |
| 2023 | January | 22 | | | | | [22] |
| 2023 | January | 22 | | | | 35 | [22] |
| 2023 | June | 22 | | | | | [22] |
| 2023 | March | 22 | 6 F | | | | [22] |
| 2023 | April | 24 | | | | 35 | [22] |
| 2023 | May | 24 | F | 150 | | | [22] |
| 2023 | May | 22 | | 38 | | | [22] |
| 2023 | May | 22 | M | 57 | | 17 | [22] |
| 2023 | May | 22 | F | 77 | | 111 | [22] |
| 2023 | January | 22 | | 72 | | 60 | [22] |

**Table 2.**
*The historical and recent records of the* Squatina Squatina *off the Mediterranean Sea.*

| Year | Month | GFCM sub-area | Sex | Total length (cm) | Gears | Depth | References |
|---|---|---|---|---|---|---|---|
| 1997 | May | 24 | | 79,89 | Bottom trawl | 120–200 | [34] |
| 2001 | Mai | 14 | M | 149 | | | [11] |
| 2001 | January | 14 | M | 149 | Bottom trawl | | Present work |
| 2003 | March | 14 | M | 131,5 | | | [11] |
| 2003 | March | 14 | M | 131,5 | Bottom trawl | | Present work |
| 2004 | August | 14 | F | 156,5 | Bottom trawl | | Present work |
| 2004 | August | 14 | F | 123 | Bottom trawl | | Present work |

| Year | Month | GFCM sub-area | Sex | Total length (cm) | Gears | Depth | References |
|---|---|---|---|---|---|---|---|
| 2004 | July | 13 | M | 152 | Gillnets | | Present work |
| 2004 | July | 13 | M | 155 | Gillnets | | Present work |
| 2004 | September | 12 | F | 128 | Bottom trawl | | Present work |
| 2004 | September | 12 | M | 118 | Bottom trawl | | Present work |
| 2004 | September | 12 | F | 103,5 | Bottom trawl | | Present work |
| 2004 | May | 22 | F | 143,5 | Local fisherman | 64 | [14] |
| 2005 | April | 22 | M | 96,5 | Bottom trawl | 130 | [35] |
| 2006 | May | 22 | F | 104,5 | Bottom trawl | 65–75 | [14] |
| 2007 | May | 12 | M | >120 | Bottom trawl | 120 | [10] |
| 2007 | June | 12 | F | >130 | Bottom trawl | 75 | [10] |
| 2009 | August | 15 | F | 128 | Bottom trawl | | [10] |
| 2009 | February | 12 | M | >120 | longline | 70 | [10] |
| 2009 | January–June | 14 | F | 85 | Bottom trawl | | [16] |
| 2009 | January–June | 14 | F | 98 | Bottom trawl | | [16] |
| 2009 | January–June | 14 | F | 115 | Bottom trawl | | [16] |
| 2009 | January–June | 14 | F | 128 | Bottom trawl | | [16] |
| 2009 | January–June | 14 | F | 138 | Bottom trawl | | [16] |
| 2009 | January–June | 14 | F | 142 | Bottom trawl | | [16] |
| 2009 | January–June | 14 | F | 166 | Bottom trawl | | [16] |
| 2010 | March | 4 | F | 120 | Bottom trawl | 100 | [36] |
| 2011 | June | 15 | F | 49,5 | Bottom trawl | 105 | [10] |
| 2012 | June | 12 | M | >120 | Longline | 50 | [10] |
| 2013 | November | 14 | M | 58 | Bottom trawl | | Present work |
| 2013 | March | 12 | M | >120 | Longline | 100 | [10] |
| 2014 | June | 24 | | 69 | | 513 | [37] |
| 2014 | December | 15 | F | 43 | Bottom trawl | 137 | [10] |
| 2015 | July | 14 | F | | Gillnets | | Present work |
| 2015 | December | 13 | F | 140 | | | Present work |
| 2015 | June | 22 | | 37,4 | Bottom trawl | 415–430 | [5] |
| 2015 | | 22 | | | | | [5] |
| 2015–2016 | | 22–23 | | | | | [5] |
| 2015–2016 | | 22–23 | | | | | [5] |
| 2015–2016 | | 22–23 | | | | | [5] |
| 2015–2016 | | 22–23 | | | | | [5] |

| Year | Month | GFCM sub-area | Sex | Total length (cm) | Gears | Depth | References |
|---|---|---|---|---|---|---|---|
| 2015–2016 | | 22–23 | | | | | [5] |
| 2015–2016 | | 22–23 | | | | | [5] |
| 2015–2016 | | 22–23 | | | | | [5] |
| 2015–2016 | | 22–23 | | | | | [5] |
| 2015–2016 | | 22–23 | | | | | [5] |
| 2016 | December | 4 | F | 130 | Bottom trawl | 100 | [36] |
| 2016 | | 22 | | | | | [5] |
| 2016 | June | 14 | M | | | | Present work |
| 2016 | June | 14 | F | 120 | | | Present work |
| 2016 | August | 15 | F | 29 | Bottom trawl | 128 | [10] |
| 2017 | February | 13 | F | 115 | | | Present work |
| 2017 | December | 16 | | | Trammel net | 70 | [19] |
| 2018 | February | 22 | | 120 | Bottom trawl | 165 | [19] |
| 2018 | April | 22 | | 100 | Bottom trawl | 130 | [19] |
| 2018 | August | 22 | | 100 | | | [22] |
| 2018 | November | 22 | | | Recreational fishing | 330 | [19] |
| 2019 | | 22 | | | | | [5] |
| 2019 | February | 24 | M | 117 | Trammel net | 47 | [38] |
| 2020 | | 22 | | | | | [5] |
| 2020 | | 22 | | | | | [5] |
| 2020 | | 22 | | | | | [5] |
| 2020 | | | | | | | [23] |
| 2021 | February | 13 | F | 150 | Bottom trawl | 90 | [10] |
| 2021 | April | 26 | F | 113,7 | Bottom trawl | 55–70 | [10] |
| 2021 | April | 22 | F | >130 | Bottom trawl | 530 | [10] |
| 2021 | April | 22 | M | 58 | | | [22] |
| 2021 | April | 16 | M | 150 | Trammel net | 34 | [10] |
| 2021 | May | 22 | F | 130 | Trammel net | 80 | [10] |
| 2021 | May | 22 | F | 120 | Trammel net | 81 | [10] |
| 2021 | May | 26 | F | 89 | Bottom trawl | 89 | [10] |
| 2021 | May | 11.2 | M | 150 | Trammel net | 40 | [10] |
| 2021 | February | 24 | | | | | [22] |

*DOI: http://dx.doi.org/10.5772/intechopen.1008752*

| Year | Month | GFCM sub-area | Sex | Total length (cm) | Gears | Depth | References |
|---|---|---|---|---|---|---|---|
| 2021 | November | 24 | | | | | [22] |
| 2021 | July | 13 | | 150 | | | Present work |
| 2021 | April | 13 | | 160 | | | Present work |
| 2021 | April | 12 | M | 140 | Bottom trawl | | Present work |
| 2022 | September | 22 | F | 120 | Bottom trawl | | [10] |
| 2022 | February | 12 | M | 135,5 | Bottom trawl | | [24] |
| 2022 | February | 12 | M | 140 | Bottom trawl | | [24] |
| 2022 | February | 12 | M | 1350 | Bottom trawl | | Present work |
| 2023 | May | 24 | F | | | | [22] |
| 2024 | May | 13 | M | 140 | Bottom trawl | | Present work |
| 2024 | May | 13 | M | 138 | Bottom trawl | | Present work |

**Table 3.**
*The historical and recent records of the* Squatina aculeata *off the Mediterranean sea.*

## 4. Diversity and status

Angel sharks occur worldwide in temperate and tropical waters [39]. They usually inhabit sandy or muddy bottom close to 150 m until 500 m in depth. They are known to entomb their body in seabeds during the day, where they remain camouflaged and take a more active approach at night [40].

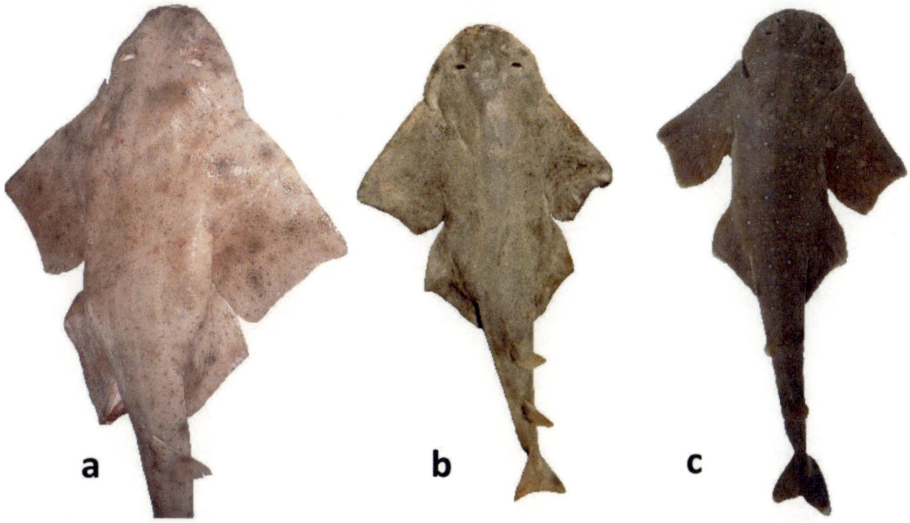


**Figure 1.**
*The three angelsharks species recorded in the Mediterranean Sea: a:* Squatina aculeate; b: Squatina squatina and c: Squatina oculata.

In the Mediterranean Sea, three species of angel shark are present with overlapping ranges: the Smoothback angelshark *Squatina oculata*, Bonaparte, 1840, the sawback angelshark *Squatina aculeata*, Cuvier, 1829 and the common angelshark *Squatina squatina* (Linnaeus, 1758) (**Figure 1**).

Globally, Squatinidae population is estimated to have been reduced by at least 90% over the last 45 years, based on population declines, area of occupancy and the contemporary rarity of these species [41].

All three angel sharks species are classified as critically endangered (CR) in the red list of the IUCN due to the high risk of extinction in the wild [42–44]. They are also listed in Annex II of the SPA/BD Protocol (Specially Protected Areas and Biological Diversity in the Mediterranean) of the Barcelona Convention (BCN Conv.). Moreover, the common angel shark *Squatian squatina* is registered in Appendix I (Endangered migratory species) of the Convention on Migratory Species (CMS).

## 5. Angelsharks statistic

Although angel shark species are protected and their capture is prohibited. Tunisia and France report their official production. Squatinidae contributed an average of 0.69% to Mediterranean cartilaginous fish production, according to FAO official statistics. A mean of 93 tons' have been landed every year during the last 22 years in the area (**Figure 2**).

These statistics are aberrant and indicate clearly that there is a misidentification with other species (e.g., guitarfishes) because of the rarity of angel sharks in the area. Hence, the improvement of species identification is necessary for statistic collectors in the Mediterranean Sea.

**Figure 2.**
*Elasmobranchs and angelsharks production according to FAO statistics from 2000 to 2022.*

## 6. Spatial occurrences

The highest number of recent angel shark sightings were recorded in the eastern Mediterranean Sea (GSA 22), followed by the central Mediterranean Sea (GSA 13, 14 and 15). No signalization of Squatinidae on the western Mediterranean Sea and the Black Sea was registered during the last 30 years.

The northeastern part of the Mediterranean basin (GSA 22) seems to be an important area for the three angel sharks species, *S. squatina, S. aculeata* and *S. oculata*, respectively, with 40%, 28% and 37% of recorded specimens (**Figure 3**). Predictive distribution maps elaborated in the Aegean Sea based on data from the Angel Shark Conservation Network and projects occurring in the area indicate that the area represents a critical habitat for the three angelsharks species [5, 19, 22, 25].

Analysis data show that among the central Mediterranean Sea, the Tunisian coasts (GSA 14, 13 and 12) are a key area for these vulnerable species. Historically, the zone shelters an important population of Squatinidae [45, 46].

The Maltese coast (GSA 15) appears to be a critical area for the smoothback angel shark *Squatina oculata* by about 18% of recorded specimens.

Many angel sharks sightings have been also documented during fish market surveys in Libya (GSA 21). It is common to see angel sharks landed and sold year-round, indicating the importance of their population in the area [23]. Although Squatinidae studies in the area are scarce, there is some anecdotal evidence of higher angel shark catch between December and April.

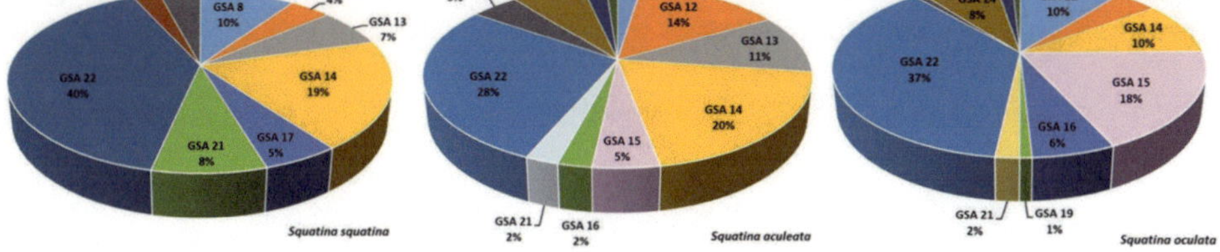


**Figure 3.**
*Distribution of Squatinidae sighting according to GFCM Subarea. a:* S. squatina*; b: S.* aculeata *and c.* S. oculata.

## 7. Temporal occurrences

Season of capture data were reported for 69 *S. squatina*, 68 S. *aculeata* and 112 *S. oculata*. The species were accidentally captured all over the year, with significant differences between seasons (**Figure 4**). The greatest number of observations was recorded in spring for *S. squatina* and *S. aculeata* and in winter for *S. oculata*.

Like the majority of elasmobranch species, angel sharks get closer to the coast for reproduction and parturition during spring [47]. This preference makes the interaction with gears and the bycatch more probable.

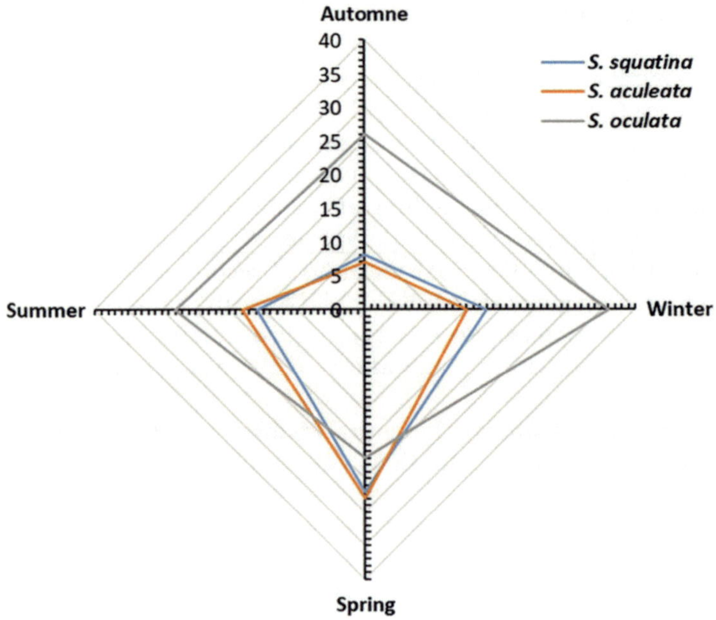


**Figure 4.**
*Seasonal sighting of Squatinidae in the Mediterranean Sea.*

## 8. Gear interactions

Along the Mediterranean coasts, angel sharks are captured as incidental catch. The bottom trawl and the trammel nets were the most gear affecting the Squatinidae population with 50% and 37% of the captured specimens, respectively (**Figure 5**). Angel sharks are highly sensitive to these gears due to their benthic and shallow habitat range

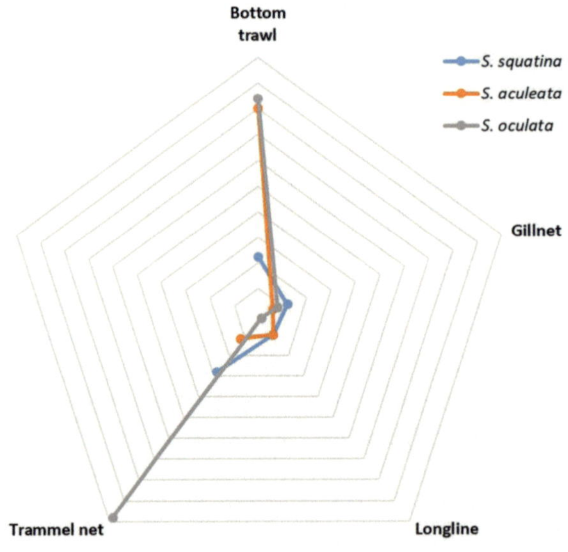


**Figure 5.**
*Distribution of angel sharks captured according to fishing gear in the Mediterranean Sea.*

and the intense fishing activity in the area. Small-scale vessels using nets with trawlers account for about 90% of the fishing fleet operating the Mediterranean Sea [9].

In the eastern Mediterranean Sea, nets represent more gears, causing the capture of angel sharks; however, most specimens bycatches were alive [22], and the possibility of their safety release is important. The bottom trawling was listed as the biggest threat for these taxa in the central Mediterranean Sea [10]; however, no data on the percentage of mortalities caused by this gear is available.

## 9. Size and sex occurrences

All Squatinidae life stages were reported in the Mediterranean Sea during the last 30 years. Juveniles and large specimens are still occurring in the area. However, the maturing specimens dominate the capture for all angel sharks species (**Figure 6a**).

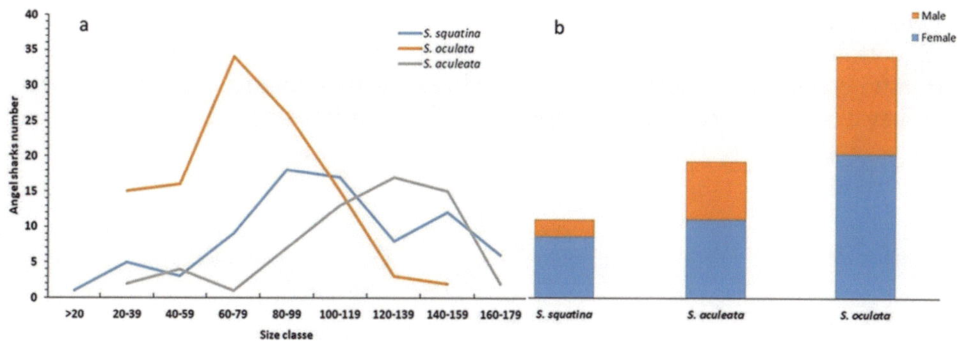


**Figure 6.**
*Size (a) and sex occurrences (b) of Squatinidae in the Mediterranean Sea.*

**Figure 7.**
*Pregnant females captured accidentally in the Mediterranean Sea. a:* Squatina oculata *(photo E. Özgür Özbek); b.* Squatina squatina. *(Photo Kabaskal H.).*

The rates of femininity for angel sharks species examined are respectively 78.13, 57.14 and 59.60% for *S. squatina*, *S. aculeate* and *S. oculata* (**Figure 6b**). Consequently, females outnumbered the males by 3.57, 1.33 and 1.48: 1, respectively, for the three species, which were a significant departure from the hypothetical 1:1.

These results, coupled with the presence of pregnant females with full-term fetuses and newborns in landing [17, 48], represent a great loss in population recruitment (**Figure 7**).

## 10. Bathymetric occurrence

Squatinidae data records prove that the common angelshark *S. squatina* is found at depths from 5 to 236 m but is most common in shallow water under 50 m. However, the Smoothback angelshark *S. oculata* is more observed at a depth above 100 m. The sawback angelshark *S. aculeata* can be found up to 530 m, but it is more abundant between 50 and 100 m depth (**Figure 8**).

Squatinidae occupied the continental shelve to the uppermost slope habitat on soft, muddy bottoms. More investigation in life stage bathymetric repartition represents key data to understanding where to look for their important areas.

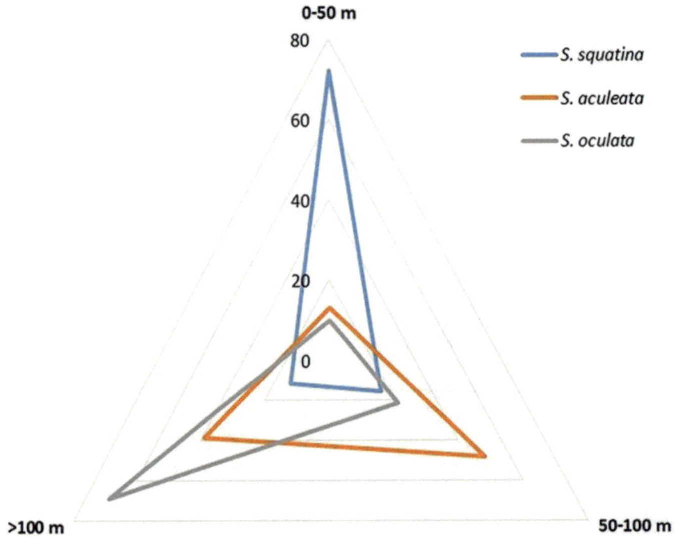


**Figure 8.**
*Bathymetric occurrences of Squatinidae in the Mediterranean Sea.*

## 11. Critical area

Gathering data on captured or observed water angelshark in the Mediterranean Sea allowed us to identify a potential critical area for these species (**Figure 9**). The GFCM sub-area 12, 13, 14 and 15 in the central Mediterranean Sea and the GSA 21, 22, 23 and 24 in the Eastern represent a possible important area for Squatinidae species.

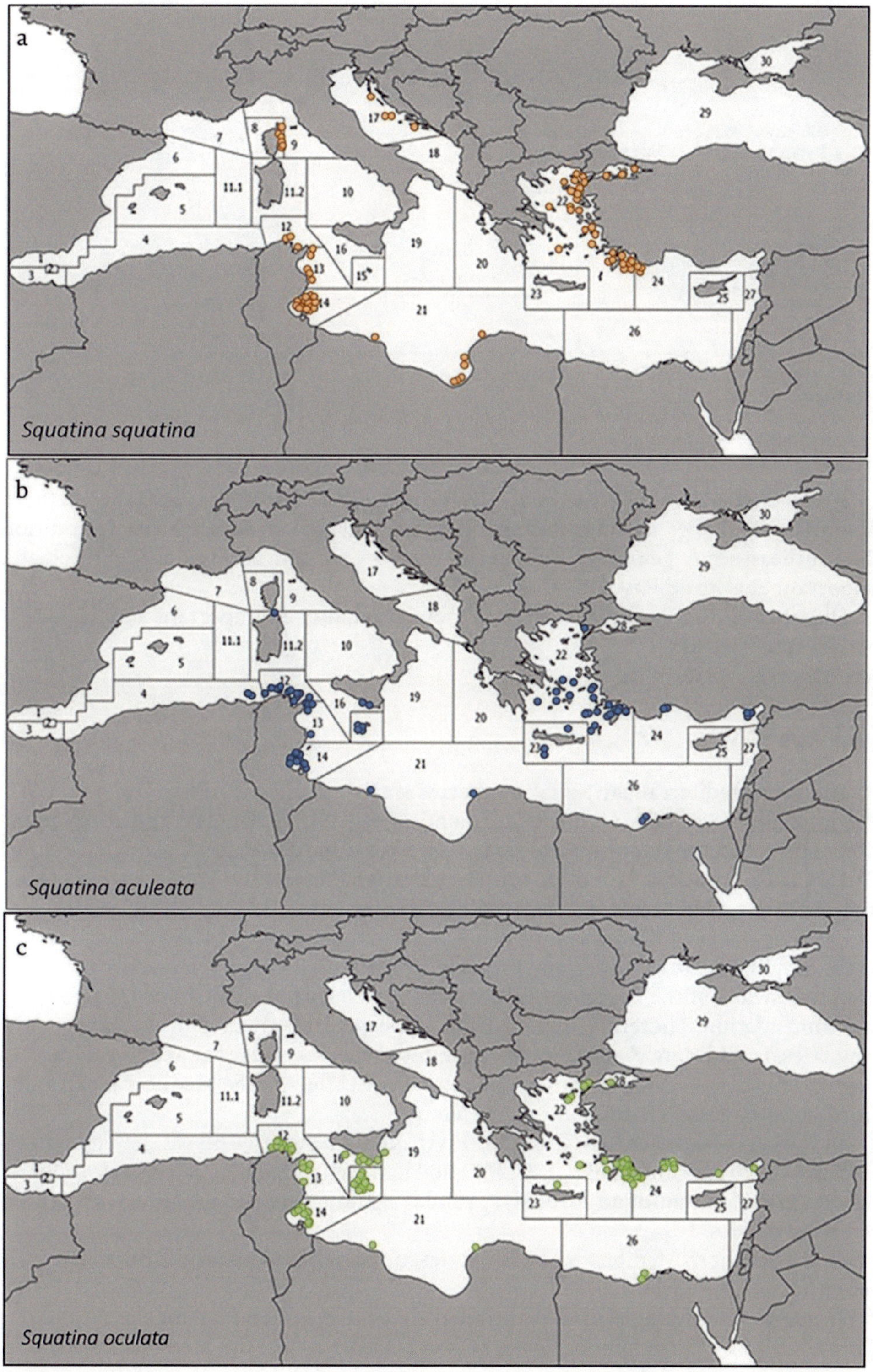


**Figure 9.**
*Habitat maps angelsharks species in the Mediterranean Sea. a:* S. squatina*; b:* S. aculeata*; c:* S. oculata.

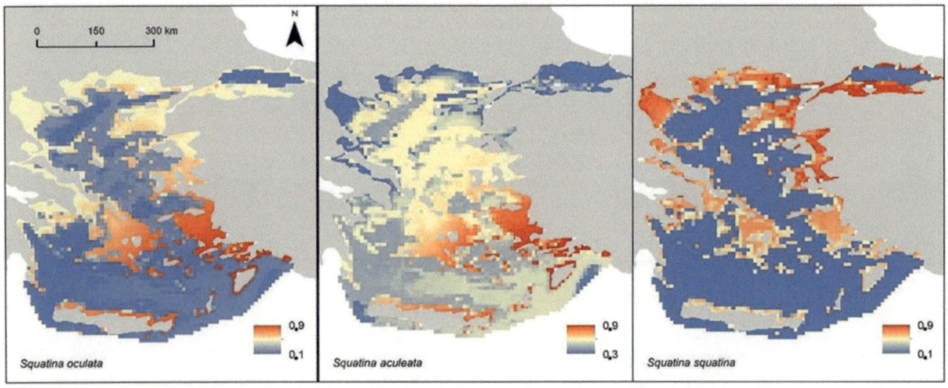


**Figure 10.**
*Habitat suitability maps for the three angel shark species in GSAs 22, 23 and 28, with red being areas that are highly suitable and blue unsuitable [5].*

Alyan et al. [22] and Giovos et al. [19] provided further support that the northern Aegean Sea housings are important habitats for angel sharks (**Figure 10**). In addition, the Southeastern Aegean Sea (located between Greece and Türkiye) is identified as an Important Shark and Ray Area (ISRA) [49].

Historically, the Tunisian coasts are known to shelter an important Squatinidae population [45, 46].

## 12. Legislation

All three Mediterranean Squatina species are listed under binding Recommendation GFCM/42/2018/2 (amending GFCM/36/2012/3). This recommendation prohibits the retention and sale of 24 elasmobranchs listed in Annex II of the Barcelona Convention. This regulation is operational in all GFCM application areas and requires transposition into national legislation.

The Regulation (EU) 2015/2102 prohibits also the retention of all species listed in the SPA/BD Protocol (Specially Protected Areas and Biological Diversity in the Mediterranean) of the Barcelona Convention in line with the GFCM/36/2012/3 recommendation. Therefore, all Squatinidae species are protected in the European countries. In addition, *S. squatina* is a prohibited species under the Technical Measure, Regulation (EU) 2019/1241, which applies to the EU fleet in the Mediterranean and third-country vessels fishing in Union waters.

In Turkey—Fisheries Law (No: 1380) is the main legislative instrument governing fisheries. In 2018, Communique 2018/19 updated Article 5 of the Turkish Prohibited Species lists (Communique 2016/35), prohibiting targeting and retention of all three Squatina species.

In the Southern Mediterranean countries, no national regulation banned the retention of angelsharks.

In addition to the legislation mentioned above, the Action Plan for the Conservation of Cartilaginous Fishes (Chondrichthyans) in the Mediterranean Sea provides a framework for species conservation and habitat protection [50].

Furthermore, the Mediterranean Angel Sharks: Regional Action Plan (RAP) developed a road map for the regional governments to use for preserving angel sharks [41].

The objective is to restore robust populations capable of fulfilling their ecological roles in equilibrate ecosystems *via* sub-regional action plans. Up to now, sub-regional Action Plans for the Ionian Sea, Aegean Sea/Crete, Northern Levant Sea and Cyprus GSAs 21, 22, 23, 24, and 25 have been developed [23, 51–54]. They detailed guidance to improve the angelsharks needs to help their population recruitment. The sub-regional action plan in GSA 12, 13 and 14 are ongoing.

These action plans are the results of work funded and supported by the Shark Trust.

## 13. Threats

Most threats of the majority of taxa in the Mediterranean Sea are grouped under biological resource use related to fishing activities. This is conventional, given that the basin has been subject to intense fishing activity since ancient times [9].

Priority threats of angel sharks remain largely the same across the 22 Mediterranean countries. They include the negative effects of differing fishing gears, habitat degradation (due essentially to the trawling effect on the seabed), illegal and unregulated fishing, the lack of knowledge regarding their habitat preference and the impacts of anthropogenic disturbances. The absence of species-specific landings and misidentification issues in both small and large-scale fisheries represent a constraint that compounds the threats [39].

## 14. Recommendation and conclusions

In light of the critical situation and the high risks of extinction of angel sharks in the Mediterranean Sea, effective protective measures across the area should be adopted. These measures must be prioritized in fisheries, habitats and legislation. Developing mitigation measures to minimized mortality, identification of critical habitats (CASAs) and establishment, implementing and enforcement of national and regional legislation represent a key element to improve the management of the species.

Considering the rarity of angel sharks in the area and the scarcity of biological and habitat use data, it is basic to implicate the fishermen's community in the conservation vision. Developing training programs to educate fishers about the status and regulation of angel sharks in the Mediterranean Sea, as well as the best handling techniques and best practices, are crucial to limit the impacts of fisheries and help rebuild their populations.

## Author details

Samira Enajjar[1*], Bechir Saidi[2] and Mohamed Nejmeddine Bradai[3]

1 National Institute of Science and Technology of the Sea, Marine Biodiversity Laboratory, Carthage University, Sfax, Tunisia

2 Faculty of Sciences and Techniques of Sidi Bouzid, Kairouan University, Sidi Bouzid, Tunisia

3 Faculté des Sciences de Sfax, ASCOB-Syrtis, Sfax, Tunisia

*Address all correspondence to: enajjarsamira@yahoo.fr

## References

[1] Dulvy NK, Allen DJ, Ralph GM, Walls RHL. The Conservation Status of Sharks, Rays and Chimaeras in the Mediterranean Sea [Brochure]. Malaga, Spain: IUCN; 2016

[2] Ferretti F, Myers RA, Serena F, Lotze HK. Loss of large predatory sharks from the Mediterranean Sea. Conservation Biology. 2008;**22**:952-964

[3] Bradai MN, Saidi B, Enajjar S. Overview on Mediterranean shark's fisheries: Impact on the biodiversity. In: Türkoğlu M, Önal U, Ismen A, editors. Marine Ecology Biotic and Abiotic Interactions. London, UK: IntechOpen; 2018. pp. 211-230. DOI: 10.5772/74923

[4] Bradai MN, Enajjar S, Saidi B. Sharks' status in the Mediterranean sea urgent awareness is needed. In: Bradai MN, Enajjar S, Saidi B, editors. Sharks - Past, Present and Future. London, UK: IntechOpen; 2023. pp. 99-116. DOI: 10.5772/intechopen.108162

[5] Giovos I, Katsada D, Spyridopoulou RNA, Poursanidis D, Doxa A, Katsanevakis S, et al. Strengthening angel shark conservation in the Northeastern Mediterranean sea. Journal of Marine Science and Engineering. 2022;**10**:269. DOI: 10.3390/jmse10020269

[6] Pinardi N, Arneri E, Crise A, Ravaioli M, Zavatarelli M. The Physical, Sedimentary and Ecological Structure and Variability of Shelf Areas in the Mediterranean Sea. Vol. 14. Cambridge: Harvard University Press; 2006. pp. 1243-1330

[7] UNEP-MAP-RAC/SPA. In: Würtz M, Artescienza SAS, Cebrian D, Requena S, editors. Sicily Channel/Tunisian Plateau: Topography, Circulation and their Effects on Biological Components. Tunis: RAC/SPA; 2015; 39 pp

[8] Emig C, Geistdoerfer P. The Mediterranean deep-sea fauna: Historical evolution, bathymetric variations and geographical changes. Carnets de Géologie, Notebooks on Geology. 2004;**1**:10

[9] FAO. The State of Mediterranean and Black Sea Fisheries 2023 – Special Edition. Rome: General Fisheries Commission for the Mediterranean; 2023. DOI: 10.4060/cc8888en

[10] Zava B, Insacco G, Deidun A, Said A, Ben Souissi J, Nour OM, et al. Records of the critically endangered *Squatina aculeata* and *Squatina oculata* (*Elasmobranchii*: *Squatiniformes*: *Squatinidae*) from the Mediterranean Sea. Acta Ichthyologica et Piscatoria. 2022;**52**:285-297

[11] Bradai MN, Saidi B, Ghorbel M, Bouain A, Guelorget O, Capape C. Observations sur les requins du golfe de Gabes (Tunisie meridionale, Mediterranee centrale). Mesogée. 2002;**60**:61-77

[12] Kabasakal H, Kabasakal E. Sharks captured by commercial fishing vessels off the coast of Turkey in the northern Aegean Sea. Annales, Series Historia Naturalis Archives. 2004;**14**:171-180

[13] De Maddalena A, Baensch H, Heim W. Sharks of the Mediterranean: An Illustrated Study of all Species. Jefferson, NC: McFarland & Co.; 2016

[14] Corsini M, ZAVA B. Recent capture of *Squatina oculata* and *Squatina aculeata* from Dodecanese Islands (SE Aegean Sea, Eastern Mediterranean).

Biologia Marina Mediterranea. 2007;**14**(2):352-353

[15] Zava B, Fiorentino F, Serena F. Occurrence of juveniles *Squatina oculata* Bonaparte, 1840 (*Elasmobranchii: Squatinidae*) in the strait of Sicily (Central Mediterranean). Cybium. 2016;**40**(4):341-343. DOI: 10.26028/cybium/2016-404-011

[16] Hamdaoui B. Les élasmobranches dans les débarquements des chalutiers au port de pêche de Sfax, golfe de Gabès [thesis]. Sfax: Sfax University; 2009

[17] Özgür Özbek E, Kabasakal H. Notes on smoothback angel shark, *Squatina oculata* (*Squatiniformes*: *Squatinidae*) caught in the Gulf of Antalya. Annales Series Historia Naturalis. 2022;**32**(1):9-16. DOI: 10.19233/ASHN.2022.02

[18] Ergüden D, Ayas D, Gürlek M, Karan S, Turan C. First documented smoothback angelshark *Squatina oculata* Bonaparte, 1840 from the North-Eastern Mediterranean Sea, Turkey. Cahiers de Biologie Marine. 2019;**60**:189-194

[19] Giovos I, Stoilas VO, Al-Mabruk SA, Doumpas N, Marakis P, Maximiadi Moutopoulos MD, et al. Integrating local ecological knowledge, citizen science and long-term historical data for endangered species conservation: Additional records of angel sharks (*Chondrichthyes*: *Squatinidae*) in the Mediterranean Sea. Aquatic Conservation. 2019;**29**:881-890

[20] Rafrafi-Nouira S, Diatta Y, Diaby A, Capapé C. Additional records of rare sharks from Northern Tunisia (Central Mediterranean Sea). Annales Series Historia Naturalis. 2019;**29**(1):25-34. DOI: 10.19233/ASHN.2019.03

[21] Yigin CÇ, Ismen A, Daban B, Cabbar K, Önal U. Recent findings of rare sharks, *Squatina oculata* Bonaparte, 1840 and *Squatina squatina* (Linnaeus, 1758) from Gökçeada Island, Northern Aegean Sea, Turkey. Journal of the Black Sea / Mediterranean Environment. 2019;**25**:305-314

[22] Ulman A, Gordon CA, Hood AR, Warren M, Pauly D. Identifying potential critical angel shark areas in Türkiye, Eastern Mediterranean based on new records of *Squatina* spp. identified through fisher engagement. Fishes. 2024;**9**(7):270. DOI: 10.3390/fishes9070270

[23] Al Mabruk S, Barker J, Gillham R, Giovos I, Hood A, Jiménez D, et al. Mediterranean Angel Sharks: SubRegional Action Plan (SubRAP) GSA 21 (Southern Ionian Sea). Plymouth, UK: The Shark Trust; 2021

[24] Rafrafi-Nouira S, Chérif M, Reynaud C, Capapé C. Captures of the rare smoothback angel shark *Squatina oculata* (Squatinidae) from the Tunisian coast (Central Mediterranean Sea). Thalassia Salentina. 2022;**44**:9-16. DOI: 10.1285/i15910725v44p9

[25] Kabasakal H, Kabasakal Ö. Status of angelshark, *Squatina squatina* (*Elasmobranchii*: *Squatiniformes*: *Squatinidae*) in the sea of Marmara. Annales, Series Historia Naturalis Archives. 2014;**24**:41-46. Available from: http://zdjp.si/wp-content/uploads/2015/08/kabasakal-kabasakal1.pdf

[26] Karakulak FS, Erk H, Bilgin B. Length-weight relationships for 47 coastal fish species from the Northern Aegean Sea, Turkey. Journal of Applied Ichthyology. 2006;**22**:274-278

[27] Işmen A, Çiğdem Yığın C, Altınağaç U, Ayaz A. Length-weight relationships for ten shark species from Saros Bay (North Aegean Sea).

Journal of Applied Ichthyology. 2009;**25**(suppl. 1):109-112. DOI: 10.1111/j.1439-0426-2009.01263.x

[28] Holcer D, Bojan L. New data on the occurrence of the critically endangered common angleshark, *Squatina squatina* in the Croatian Adriatic Sea. Natura Croatica. 2017;**26**(2):313-320. DOI: 10.20302/NC.2017.26.23

[29] Kabasakal H. Finally, under protection! Status of the angel shark, *Squatina squatina* (Linnaeus, 1758) in Turkish seas, with notes on a recent sighting and incidental captures. Annales, Series Historia Naturalis. 2019;**29**:17-24

[30] Akyol O, Ünal V, Capapé C. Occurrence and biological observations on angel shark *Squatina squatina* (*Chondrichthyes*: *Squatinidae*) from the Turkish waters (Eastern Mediterranean). Turkish Journal of Fisheries and Aquatic Sciences. 2015;**15**:951-955

[31] Lapinski M, Giovos I. New records of the critically endangered *Squatina squatina* (Linnaeus, 1758) from Corsica, France. Acta Adriatica. 2019;**60**(2):205-210

[32] Zammit CM. Les élasmobranches, les tortues marines et les cétacés dans les pêcheries de Teboulba [thesis]. Sfax: Sfax University; 2020

[33] Rafrafi-Nouira S, Cherif M, Reynaud C, Capapé C. Captures of common angelshark *Squatina squatina* (Squatinidae) from the northern Tunisian coast (Central Mediterranean Sea). Cahiers de Biologie Marine. 2023;**64**(3):203-207. DOI: 10.21411/cbm.a.7e50396f

[34] Basusta N. Occurrence of a Sawback Angelshark (*Squatina aculeata* Cuvier, 1829) off the eastern Mediterranean coast of Türkiye. Turkish Journal of Veterinary and Animal Sciences. 2002;**26**:1177-1179. Available from: https://dergipark.org.tr/en/pub/tbtkveterinary/issue/12560/151877

[35] Filiz H, Irmak E, Mater S. Occurrence of *Squatina aculeata* Cuvier, 1829 (*Elasmobranchii*, *Squatinidae*) from the Aegean Sea, Türkiye. Ege Journal of Fisheries and Aquatic Sciences. 2005;**22**:451-452

[36] Hemida F, Reynaud C, Capapé C. First Records of Sawback Angelsharks *Squatina aculeata* (*Squatinidae*) from the Algerian coast (Southwestern Mediterranean Sea). Annales, Series Historia Naturalis Archives. 2024;**34**:21-26p

[37] Ergüden D, Bayhan KY. Three fish species known to be rare for Türkiye, captured from the Northeastern Mediterranean coast of Türkiye, Mersin Bay, Sudis hyalina Rafinesque, Chlopsis bicolor Rafinesque, *Squatina aculeata* Cuvier. International Journal of Scientific and Technology Research. 2015;**1**:1-8

[38] Ergenler A, Turan F, Turan C. Occurrence of a sawback angelshark (*Squatina aculeata* Cuvier, 1829) from the Iskenderun Bay, North-Eastern Mediterranean coast of Turkey. Biharean Biologist. 2020;**14**:57-59

[39] Compagno LJV. FAO species catalogue. Vol. 4. Sharks of the world. An annotated and illustrated catalogue of shark's species known to date. In: Hexanchiformes to Lamniformes. Vol. 4. Rome: FAO Fisheries Synopsis; 1984. p. 125, Part 1, 258 p

[40] Ebert DA, Dando M. Field Guide to Sharks, Rays & Chimaeras of Europe and the Mediterranean. United Kingdom: Princeton University Press; 2021. DOI: 10.2307/j.ctv12sdwkk

[41] Gordon CA, Hood AR, Al Mabruk S, Barker AA, Bartolí A, Ben AS, et al. Mediterranean Angel Sharks: Regional Action Plan. United Kingdom: The Shark Trust; 2019. 36 p

[42] Morey G, Barker J, Bartolí A, Gordon C, Hood A, Jimenez-Alvarado D, et al. *Squatina aculeata*. In: The IUCN Red List of Threatened Species. UK: IUCN; 2019. DOI: 10.2305/IUCN

[43] Morey G, Barker J, Bartolí A, Gordon C, Hood A, Meyers EKM, et al. *Squatina oculata*. In: The IUCN Red List of Threatened Species. Gland, Switzerland: IUCN; 2019. DOI: 10.2305/IUCN.UK.2019-1

[44] Morey G, Barker J, Hood A, Gordon C, Bartolí A, Meyers EKM, et al. *Squatina squatina*. In: The IUCN Red List of Threatened Species. Gland, Switzerland: IUCN; 2019. DOI: 10.2305/IUCN.UK.2019-1.RLTS.T39332A117498371.en

[45] Capapé C, Quignard JP, Mellinger J. Reproduction and development of two angel sharks, *Squatina squatina* and *S. oculata* (*Pisces*: *Squatinidæ*), off Tunisian coasts: Semi–delayed vitellogenesis, lack of egg capsule and lecithotrophy. Journal of Fish Biology. 1990;**37**:347-356

[46] Capapé C, Diatta Y, Seck AA, Guélorget O, Ben Souissi J, Zaouali J. Reproduction of the sawback angelshark *Squatina aculeata* (*Chondrichthyes*: *Squatinidae*) off Senegal and Tunisia. Cybium. 2005;**29**:147-157

[47] Capapé C, Tomasini JA, Quignard JP. Les elasmobranches pleurotrême de la côte du Languedoc (France méridionale): Observations biologiques et démographiques. Vie et Milieu/ Life and Environment. Observatoire Océanologique - Laboratoire Arago; 2000. pp. 123-133

[48] Kabasakal H. A preliminary social media survey of sharks and batoids captured in North Aegean Sea commercial fisheries. Annales, Series Historia Naturalis. 2023;**33**(2):165-186. DOI: 10.19233/ASHN.2023.20

[49] IUCN SSC Shark Specialist Group. South-Eastern Aegean Sea ISRA Factsheet. Dubai: IUCN SSC Shark Specialist Group; 2023

[50] SPA/RAC–UN Environment/MAP. In: Bradai MN, editor. Action Plan for the Conservation of Cartilaginous Fishes (Chondrichtyans) in the Mediterranean Sea. Tunis: SPA/RAC; 2020, 18 pp

[51] Gordon CA, Hood AR, Giovos I, Naasan Aga R, Spyridopoulou I, Ozturk A, et al. Mediterranean Angel Sharks Subregional Action Plan, Aegean Sea and Crete. Plymouth, UK: The Shark Trust; 2020

[52] Fakıoglu E, Gillham R, Hood A, Kabasakal H, Oruc A, Ozturk A, et al. A Mediterranean Angel Sharks Subregional Action Plan (SubRAP) GSA24, Northern Levant Sea. Plymouth, UK: The Shark Trust; 2021

[53] Giovos I, Gillham R, Hadjioannou L, Hood A, Kleitou P, Aga-Spyridopoulou RN, et al. Mediterranean Angel Sharks: SubRegional Action Plan (SubRAP) GSA 25* (Cyprus -Republic of Cyprus). Plymouth, UK: The Shark Trust; 2021

[54] Bengil E, Godley B, Gillham R, Hood A, Snape R. Mediterranean Angel Sharks: SubRegional Action Plan (SubRAP) GSA 25*(Northern Cyprus). Plymouth, UK: The Shark Trust; 2021

Chapter 2

# Sea Turtles in Tunisia: An Overview on their Status and Conservation Effort

*Mohamed Nejmeddine Bradai, Imed Jribi, Bechir Saidi and Samira Enajjar*

## Abstract

Three species of sea turtles occur in Tunisia; the loggerhead is the most common. Tunisian coasts are important for this species in the Mediterranean Sea; the Gulf of Gabes is considered a wintering and foraging area. Among many threats, the bycatch is the most impacting. Many studies were carried out on these species and many measures and mitigations were the monitoring of the nesting sites, the ssuch asundertaken tranding events and the development of ..centerssea turtle rescue Although sea turtles are legally protected in Tunisia, more effort to mitigate bycatch is needed. In this chapter, we focus on the compilation of main data on sea turtles and their analysis, in addition to our own new observations, to propose some recommendations for conservation.

**Keywords:** sea turtles, nesting, bycatch, mitigation, conservation effort, Tunisia

## 1. Introduction

### 1.1 Geographic location of Tunisia

Tunisia is a Mediterranean country in Northern Africa, bordering the Mediterranean Sea. Its geographic coordinates are 34°00′N 9°00′E, and it lies between latitudes 30° and 38°N and longitudes 7° and 12°E (**Figure 1**).

### 1.2 Coastal length and characteristics

The Tunisian total coastline extends for 2290 km with 1566 km of coastline, 267 km of artificialized linear (Port, Marina, etc.) and a linear of 457 km of islands, islets and archipelagos [1].

The north coasts are under the influence of the Atlantic current. The continental shelf is reduced with the presence of rocky bottoms.

The long of the eastern coasts, the bottom of the sea is homogeneous and the continental shelf is very large, especially at the Gulf of Gabes level. This region is

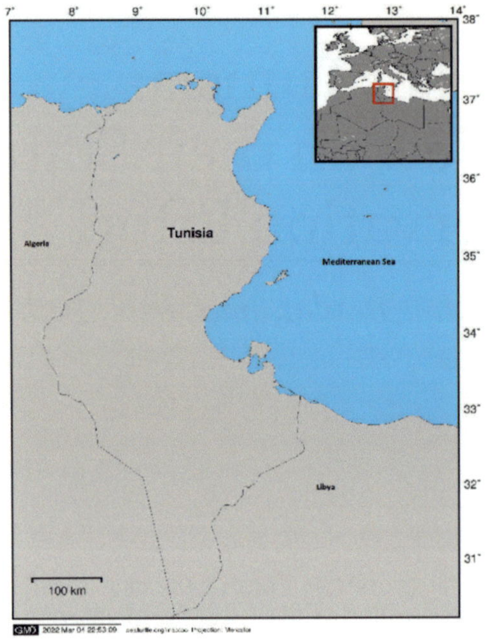


**Figure 1.**
*Map of Tunisia. (The map was created thanks to the Maptool program. Maptool is a product of SEATURTLE.ORG).*

characterized by a semi-diurnal tide with a high amplitude (until 2 m). In this sector, the Atlantic current loses its influence. The Gulf of Gabes presents hydro-dynamic and physical and chemical features different from those of the North. The temperature and the salinity are, for example, more elevated.

The important surface of the continental shelf of the Tunisian southeast coasts, the easy access to fishing zones and the presence of the *Posidonia* sea bed that constitutes nurseries for several species of vertebrates and invertebrates made this region an important fishing zone of Tunisia.

From a bio-geographic point of view, the zone Center and especially the South, which is dominated by sandy and muddy funds, have a subtropical affinity characteristic of the oriental basin.

The Tunisian coast presents several well-differentiated landscapes:

- the sandy beaches, which occupy approximately 575 km, constitute the most present landscape on the Tunisian coasts and shelter the major part of the population, essentially in the bays and gulfs of the eastern side of the country;

- the cliff coasts of about 400 km, which are found mainly on the northern facade and to the east of the main capes;

- the coastline with sandy dunes, covered with forest plantations (Zoueraa, Bizerte, Gammarth, Oued Abid, Medfoun, Ghedhabna);

- wetlands.

*DOI: http://dx.doi.org/10.5772/intechopen.1008239*

| | North (GSA 12) | Center (GSA 13) | South (GSA 14) | Total |
|---|---|---|---|---|
| Trawlers | 48 | 98 | 263 | 409 |
| Small-scale vessels (With and without engine) including mainly bottom and surface longlines and many kinds of nets. | 2272 | 2745 | 6451 | 11,468 |
| Purse-seiners | 41 | 198 | 151 | 390 |
| Total | 2361 | 3041 | 6865 | 12,267 |

**Table 1.**
*Fishing effort (total number of vessels in 2022).*

### 1.3 Fishing activities

The Tunisian port chain is made up of 40 ports: 10 deep-sea ports sheltering boats intended for trawling, tuna purse seine and coastal fishing, 22 coastal ports and 8 landing sites: Ten ports north of the country (GSA12: Northern Tunisia), 10 ports in the East (GSA13: Gulf of Hammamet) and twenty ports in the Gulf of Gabes region (GSA 14). (GSA: Geographic sub-area according to the General Fishing Commission of the Mediterranean Sea (GFCM)).

The maritime fishing units active in Tunisia number 12,267 including 11,468 small-scale vessels (**Table 1**) generating a maritime population of around 100,000 fishermen and seafarers.

Several types of fishing are practiced in Tunisia: bottom trawl fishing, purse seine fishing small and large pelagic fish. The smallscale fishery is characterized by a diversification of fishing gears and consequently of metiers. The main ones are: Trammel nets, gillnets, bottom and surface longline.

## 2. Sea turtles in Tunisia

### 2.1 Sea turtle diversity

Three species of sea turtles are known in Tunisia; the loggerhead turtle *Caretta caretta*, the green turtle *Chelonia mydas* and the leatherback turtle *Dermochelys coriacea*. The first species is common and nests on some beaches. The green turtle is rarely reported. The leatherback turtle is regularly observed [2].

The Tunisian coasts, mainly the Gulf of Gabès, are of capital importance for the populations of sea turtles in the Mediterranean. The Gulf of Gabès is considered a wintering and foraging area for *C. caretta* [2–10]. This importance has been confirmed by recaptures of tagged female loggerheads after laying on nesting sites in Greece or of ringed juveniles and subadults at sea in the northern Mediterranean. Similarly, satellite monitoring has confirmed such migrations. This migratory gathering is explained by the North-South thermal gradient of the surface waters. The turtles would seek warmer waters. The second reason could be trophic. Turtles feed in winter in southern Tunisia, they eat mainly benthic invertebrates (gastropods, crustaceans and sea cucumbers).

Moreover, regular nesting sites of the loggerhead *Caretta caretta* are known on Tunisian coasts (Kuriat and Chebba) and benefit from regular monitoring

**Figure 2.**
*Sea turtle carapace use (1) Cradle for babies on Kerkennah Island (2) for decoration.*

(mainly kuriat islands). Other sites of lesser importance have recently been discovered. It should also be noted that a green turtle nest was reported once on the beach of Rejich (Central East of Tunisia). This nest of the green turtle is the farthest west one discovered in the Mediterranean [11]. The most eastern nesting area occurs principally in Syria [12–14].

### 2.2 Cultural heritage of sea turtles in Tunisia

Sea turtles were in the past exploited. They are listed among captured species and in the fisheries statistics [15].

The main use of turtles before its ban was for food: "It is often brought to the markets of all the maritime towns where it is used for food" [16]. Its oil is sometimes used as medicine [17]. André [18] indicated that in the Kerkennah islands, the flesh is little sought after, but that the blood and the heart would be remedies against certain diseases. According to him, above all the carapace interests the Kerkenian to make a cradle of it (**Figure 2**). The use of carapace as a cradle is known also in Djerba island (mainly in Guellala). According to Ref. [19], dried eyes prepared as amulets or very young individuals associated with sponges in a basket hung over the head of a bed to protect fishermen from shark attacks or improve women fertility. According to Ref. [3], A survey was conducted in 1978, and it was found that sea turtles are mainly used for food and tourism (sale of shells) (**Figure 2**).

Despite its prohibition and efforts to protect sea turtles in Tunisia, the trade in a clandestine manner was reported in certain regions. A strategy to combat this phenomenon was developed in 2020 [20]. Currently, sea turtles are well protected in Tunisia.

## 3. Studies on sea turtles and effort of conservation

Scientific work on sea turtles in Tunisia began in earnest at the end of the 1980s following beach surveys which permitted to report the first nesting in 1988 on the beach located between Ras dimas and Mahdia and on Great Kuriat Island [7]. Surveys in the early 1990s also showed the importance of loggerhead turtle nesting on the Kuriat islands and on Chebba beach where 3 nests on the beach of "Sidi Messaoud" were discovered [21]. The monitoring of the nesting of the sea turtle *Caretta caretta* on the Kuriat islands and the surveys of the different coasts have shown that the Kuriat islands represent the most important nesting site in Tunisia. Scientific work on sea turtles was then diversified and touched on several themes. Several institutions and organizations have been involved in conservation efforts.

Following the ratification of many international conventions by Tunisian government (**Table 2**), many aspects of conservation were developed and made it possible to improve knowledge of this taxon.

Among the legislative conservation tools adopted by Tunisia, there is the Action Plan for the Conservation of Marine Turtles in the Mediterranean Sea (UNEP/PAM) within the framework of the Barcelona Convention and the recommendations of the GFCM and of ICCAT.

Following the ratification of the international conventions mentioned above, Tunisia has developed national legislation. The protection of sea turtles is ensured, at the national level, by the promulgation of law n°94–13 of July 31, 1994, of the Ministry of Agriculture and its implementing decree of September 28, 1995, which organizes fishing activities and the annual decree of the Ministry of Agriculture organizing hunting. These pieces of legislation prohibit the capture, peddling and trade of sea turtles.

In Tunisia, until 1989 sea turtles were sold to markets and consumed freely. After 1989 and following the ratification of international conventions and the drafting of national legislation to protect these endangered animals, such massacres are no longer seen, and turtles caught accidentally are often released at sea.

In the period 1989–2012, many efforts of conservation were undertaken:

- Launching of the monitoring of the main marine turtle nesting site in 1997;
- Launching of marine turtle's rescue center of Tunisia in 2004;
- the national stranding network, dealing with marine turtles and cetaceans, and the tissues bank of marine endangered species in 2004;

Following political and social problems that appeared in 2011–2012, little illegal trade of loggerheads was observed in some localities. Faced with this situation, Tunisia elaborated:

| Convention | Adoption | Ratification | Law n. |
|---|---|---|---|
| CITES | 1973 | 1974 | 74 - 12 of the 11/05/74 |
| Barcelona | 1976 | 1977 | 77 - 29 of the 25/05/77 |
| CMS | 1979 | 1986 | 86 - 63 of the 16/07/86 |
| Bern | 1979 | 1995 | 95 - 75 of the 07/08/95 |
| SPA Protocol new SPA Protocol (1995)<br>and its Annexes (amendment) | 1982<br>1995 | 1983<br>1998 | 83 - 44 of the 22/04/83<br>98 - 15 of the 23 /02/98 |
| CBD | 1992 | 1993 | 93 – 45 of the 03/05/93 |
| ACCOBAMS | 2001 | 2001 | 2001-68 of the 11/07/01 |

**Table 2.**
*International conventions and agreement ratified by Tunisia.*

- The National Action Plan for the Conservation of Sea Turtles

https://www.rac-spa.org/sites/default/files/doc_turtles_project/pan_totues_2020.pdf

- The national strategy to reduce the illegal trade in sea turtles

https://www.rac-spa.org/sites/default/files/doc_turtles_project/strat_tut-les_2020.pdf

Moreover, and within the framework of the implementation of the Action Plan for the Conservation of Marine Turtles in the Mediterranean Sea (UNEP/MAP) and its National Action Plan, Tunisia has undertaken various actions in the field of marine turtle conservation:

### 3.1 Sea turtles and cetaceans stranding network

The study of cetaceans and sea turtles stranded was reinforced at the beginning of 2004 by the creation of a national stranding network. This program is part of the activities of the INSTM marine biodiversity laboratories.

### 3.2 Awareness activities

Several educational activities aimed at the general public have been undertaken. The programs have mainly relied on the management of fishing activities targeted at the preservation of stocks, on the protection of threatened species and biodiversity and on the development of guidelines necessary for the management of incidentally caught endangered species. Seminars, leaflets, posters, books, radio and television broadcasts have been implemented for this purpose.

### 3.3 Monitoring of the Kuriat Islands nesting site

The beaches of Kuriat islands, which represent the most important nesting site, have been monitored since 1997 with a seasonal scientific camp. The activity responds to two concerns:

- Herpetological research;

- The conservation of sea turtles.

This monitoring is done within the framework of an annual convention between mainly the National Institute of Sciences and Technologies of the Sea (INSTM), the Coastal Protection and Layout Agency (APAL), the Special Protected Areas Regional Activity Center (SPA/RAC) and the Notre Grand Bleu association (NGB).

### 3.4 Sea turtles rescue centres

Given the importance of accidental captures of sea turtles and with the aim of helping those in difficulty, a Sea Turtle Rescue Center was created in 2004 at INSTM Monastir and a first aid centre was recently created in the Sfax Faculty of Sciences (2020). The centres contribute to the treatment of turtles stranded alive or tired after accidental captures.

### 3.5 Research activities

Several researches on sea turtles are carried out in Tunisia within the framework of the monitoring of nesting beaches and the activities of the rescue centres and the national sea turtle and cetacean stranding network. Other research activities were also carried out such as:

- Study of interactions with several fishing gears and mitigation measures to reduce bycatch;
- Studies of migration by metal tags and by satellite monitoring of turtles caught accidentally or on nesting sites;
- Genetic studies;
- Pollution by heavy metals;
- Studies on the ingestion of marine debris and particularly plastic.

## 4. Some available data on sea turtles in Tunisia

Under this subtitle, we try to compile data on some research needed for conservation programs nesting activity, bycatch and stranding.such as

### 4.1 Nesting activities in Tunisia

#### *4.1.1 Introduction*

The total length of sandy beaches for the entire Tunisian coastline is estimated at 593 km, 6% of which is bordered by a dune field [1].

An intense nesting activity of the Loggerhead turtle was reported since the mid-twentieth century on Tunisian coasts. Already in 1935, [3] wrote: "The loggerhead lays its eggs in the sand of the islands, islets and deserted shores of Tunisia as well as throughout northern Africa. Many other authors stated also that the eastern coast of Tunisia represented the most important region in North Africa for Loggerhead nesting activity, considering the immense range of uninhabited beaches [3, 22, 23].

This intense and widespread nesting of the Loggerhead turtle *Caretta caretta* mentioned in the literature along the Tunisian coastline, in particular in the southeast of the country, was not based on precise information.

Nevertheless, such activity of nesting was truly discovered for the first time in 1988 on the beach located between Ras Dimas and Mahdia and on the Great Kuriat island off Monastir [7], at Sidi Massaoud beach in Chebba [21] and at Zouaraa beach (Beja) in 2016 [24].

Currently, nesting activity of the Loggerhead turtle *Caretta caretta* occurs principally in Kuriat islands [25–30] and Chebba beaches [31–33] which are monitored. The first site is the most important in Tunisia. The two sites are located in the eastern coasts. Since 1997, an annual report is elaborated on the monitoring of the Kuriat nesting sites.

Although the smallness of the two nesting sites, Kuriat islands and Chebba, at the Mediterranean scale, the nesting activity is regularly registered, and the nests number increases since respectively 1997 and 1994.

Besides these documented loggerhead nesting sites in Tunisia, several testimonies mention the presence of other nesting sites. Inquiries about this phenomenon and an exploration of sandy beaches, along Tunisian coasts carried out in 2018 and 2019 confirmed such testimonies and discovered more nesting sites [33, 34].

An exceptional nesting event of green turtle was also recorded in the summer of 2019 in Rejich beach (Mahdia – Eastern coasts) [11, 33]. This nest represents the western most nesting record of the green turtle in the Mediterranean.

Tunisian beaches represent the westernmost nesting grounds for Loggerheads in the southern Mediterranean [7] before the new nesting site recorded recently in Algeria [35].

#### *4.1.2 Monitoring of the sea turtle Caretta caretta nesting on the Kuriat Islands*

The Kuriat or Qûrya Islands (35° 48′ 05″ N, 11° 02′ 05″ E) are two small emergences, 2 km away from each other, located east-northeast of Cape Monastir, in front of the bay of Khnis at 11 nautical miles, or about 20 km. These are two small uninhabited islets, characterized by a flat and low morphology not exceeding 4.5 m with several low-pressure areas (**Figure 3**).

The largest one, the Great Kuriat called also Qûrya El Kbira, has an ovoid shape, it is 3.5 km long by 2 km wide and covers about 270 ha.

Almost one third of the Great Kuriat shoreline is rocky and large deposits of sea grass (*Posidonia oceanica*) detritus further restrict the accessible nesting sites particularly in the south and the south-western beaches. The principal nesting beach lies on the western coast and it is almost 900 m in length.

The smallest one, Qûrya Sghira also known under the name of Cogniliera (the island of the rabbits), has an area of 50 ha, most of which is made up of flat and low

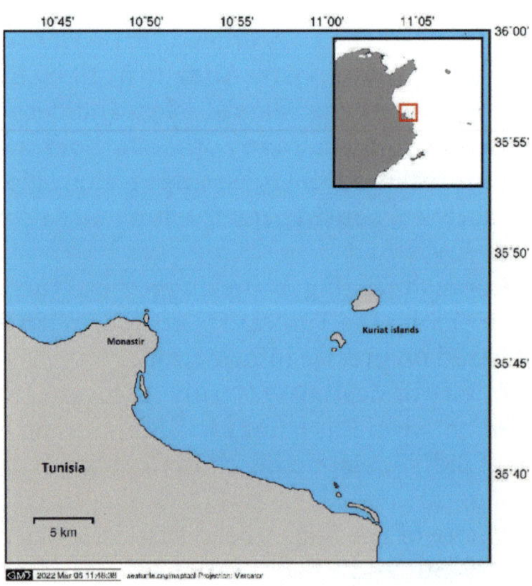


**Figure 3.**
*Geographic position of Kuriat islands off the coast of Tunisia. (The map was created thanks to the Maptool program. Maptool is a product of SEATURTLE.ORG).*

land exceedingly very rarely 0 m in the North and in the Northeast, as well as intertidal plains corresponding to the oscillation zone of the marshes. Small Kuriat has a total of 800 m of sandy beach situated in the north-eastern part of the island whereas the rest of the coastline is rocky or marshy. These two sites have been monitored regularly since 1997.

#### *4.1.2.1 Nesting period*

The monitoring of nesting on Kuriat islands since 1997 permits to locate the egg-laying period of *Caretta caretta* mainly during months of June, July and August of each year with a pic during the first half of July [29]. The laying during the month of August was registered for the first time during the nesting season of 2003 indicating a spreading of this period especially during the last years. The distribution of laying dates on the Kuriat Islands is found in the range of dates observed in the Mediterranean. Indeed, the loggerhead *Caretta Caretta* begins to lay at the end of May until the end of August, however some individuals continue to lay until early September [36]. This parameter is very important to know for the implementation of any conservation activity. Indeed, its knowledge makes it possible to reduce anthropogenic disturbances, especially when the nesting phenomenon coincides with the frequentation of nesting beaches by summer visitors and tourists [36].

The average number of nests deposited on the Kuriat Islands since 1993 is 21.4 (SD = 13.45; N = 30) (**Figure 4**). The average number of nests deposited on the Great Kuriat and on the small Kuriat are respectively 15.03 (SD = 8.87; N = 30) and 6.37 (SD = 7.02; N = 30). It should be noted that the number of nests has recorded a marked increase over the past 5 years (Mean = 45.4, SD = 3.65, N = 5), which would be the result of the protection effort deployed since the start of monitoring.

It should be noted that last 10 years nesting was recorded in May in Tunisia and in the Mediterranean countries. This early nesting activity could be related to climate change.

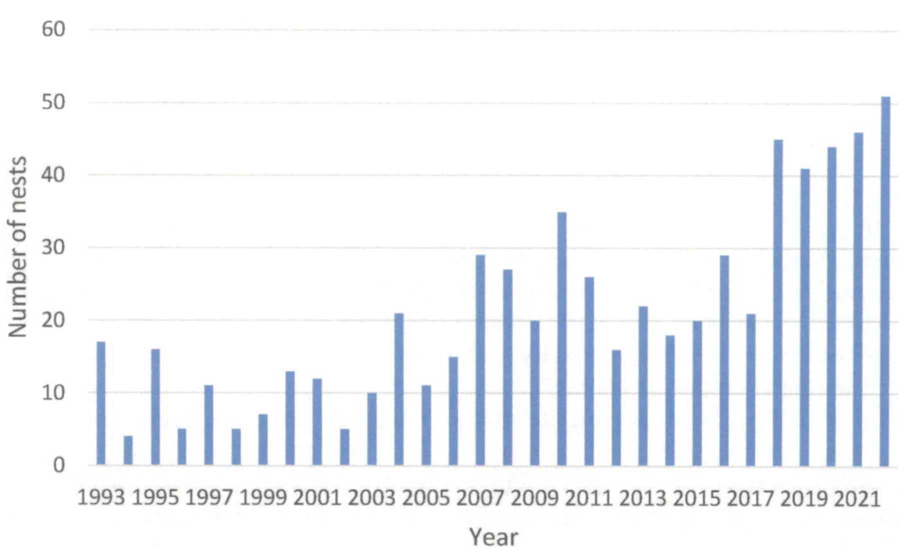


**Figure 4.**
*Number of nests yearly deposited on Kuriat islands.*

#### *4.1.2.2 Reproductive parameters*

The clutch size (the total number of eggs laid in a nest) on Kuriat Islands varies between 25 to 164 with a mean of 88.95 (N = 375). The hatching and emergence rates for nests under normal conditions exceed usually 60%, which reveals the suitability of the beaches of Kuriat islands [2, 25, 27, 29, 37].

The females' remigration interval (years between breeding migrations) on Kuriat islands is of the order of 2 years, however, intervals of 1 year have been recorded, which shows the importance of satellite monitoring to see if there are turtles who do not migrate too far to come the following year to lay. In fact, this parameter and the clutch frequency (nests per breeding season), are associated with feeding conditions and related environmental factors.

Since 1997, 81 nesting females were tagged. They had a mean CCLn-t (Curved Carapace Length) of 75.97 cm (SD = 4.13; N = 81; individual range: 68–87 cm) and a mean CCW (Curved Carapace Width) of 67.5 cm (SD = 3.86; N = 81; individual range: 61–77 cm). These data confirm that nesting females in the Mediterranean have generally curved lengths greater than 70 cm [10].

#### *4.1.2.3 Genetic*

Genetic analysis has been undertaken on the nesting site of Kuriat. Freshly dead Hatchlings of loggerhead have been analyzed for the long mtDNA control region sequences. Only the widespread Mediterranean haplotype CC-A2.1 has been detected when analyzing the long sequence of 800 bp. Anthropogenic impact linked mainly to fishing and touristic activities resulted in the observed reduced genetic diversity of the nesting population [38].

### 4.2 Monitoring of sea turtles *C. caretta* nesting sites of Chebba

Chebba is located off Cape Ras Kaboudia which is the most easterly point of the Tunisian coasts (**Figure 5**). It has the particularity to spread like a peninsula and the sea surrounds it on three sides. It has 29 km of coastline with some islets.

Two nesting sites are known in Chebba "Essir" and "Sidi Messaoud" beaches (**Figure 6**)."Essir" is the main beach of Chebba. It has a length of approximately 600 m and spreads between the two points with GPS coordinates: 35°14.386′N/011°08.557′E and 35°14.268′N/011° 08.892′E. This beach is very busy during the summer, day and night.

The beach of "Sidi Messaoud" is contrary to "Essir" and less crowded. It is a small beach of about 200 m in length located behind the fishing port and adjacent to the Roman archaeological site "Borj Khdija". It spreads between the two GPS points: 35°14.108'N/011°09.442′ E and 35°13.998′N/011°09.604′E.

The history of the nesting activity in the two nesting sites of the Chebba area can be summarized as follows (**Table 3**).

The available information on the studied nests deposited in Chebba beaches (N = 10) shows that the clutch mean is 87 eggs by nest and hatching and emergence rates are respectively 66.5 and 66.2. These rates are within the range of what is registered in the Mediterranean indicating that beaches of Chebba are suitable for nesting activity.

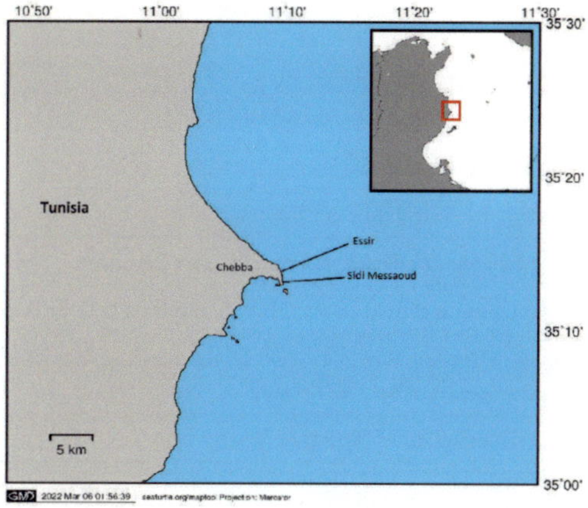


**Figure 5.**
*Geographic position of Chebba and nesting beaches.*

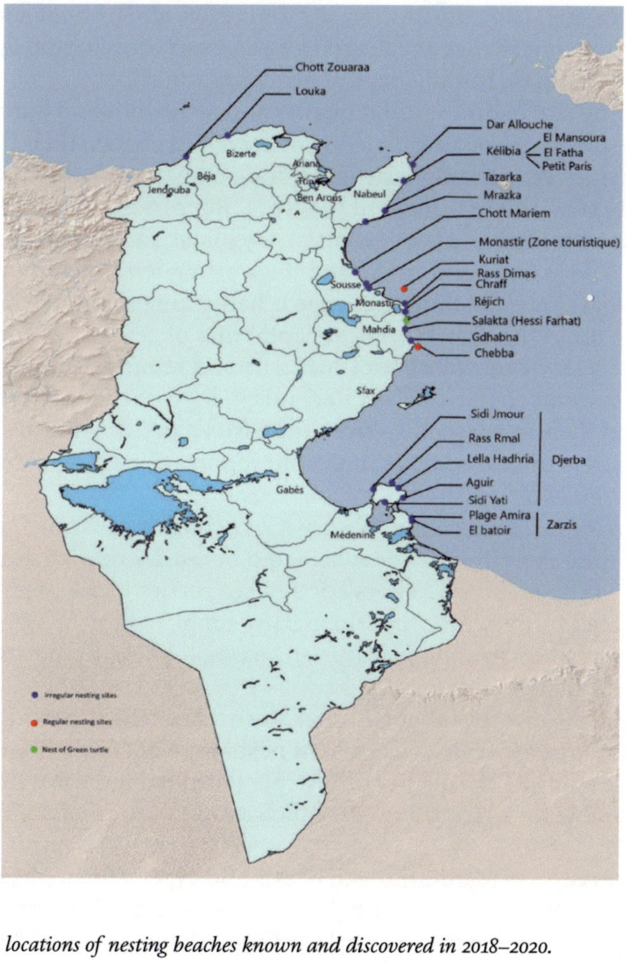


**Figure 6.**
*Map of Tunisia with locations of nesting beaches known and discovered in 2018–2020.*

| Year | Nesting activities |
|---|---|
| 1994 | First record of 2 nests of Loggerhead on Sidi Massaoud beach [21] |
| 1995 | One nest deposited on Sidi Massaoud beach [21] |
| 1996–2000 | No nesting activity recorded. |
| 2003 | Hatchlings recorded close to Essir Beach. |
| 2004–2012 | Nesting activity recorded each summer in Essir Beach. |
| 2013 | Six nests deposited at Sidi Messaoud Beach and four at Essir Beach. |
| 2019 | Four nests were deposited in 2019; two on the beach of "Essir", one at "Sidi Messaoud" and one on another beach called "El Koucha". |
| 2023 | Seven nests deposited on different beaches |

**Table 3.**
*Historic loggerhead nesting activities in Chebba sites.*

### 4.3 New nesting site discovered

We consider in this study the nesting site definition of Girondot and Fretey [39]: "A marine turtle nesting site is considered to be any surface where at least one female of any species of sea turtle has laid eggs in historical times".

Global warming is affecting habitat quality and availability on our planet and some species are predicted or already being observed to change their distribution range. Marine turtles are a particularly interesting case to study in this respect since they have already survived and adapted to several important climate change events throughout their >1 million years of evolutionary history and they colonized tropical and subtropical nesting habitats around the world notwithstanding their natal philopatry. However, current climate change is happening at a much faster rate and is expected to have profound effects on the adaptability of sea turtles whose life history is characterized by longevity, late age of maturity and temperature-dependent sex determination. It seems that in the Mediterranean, loggerhead turtles have already started to expand their nesting range from the eastern to the western basin, which has only been known to host sporadic nests but reports of nesting activity have been increasing since the 2010's [40].

In Tunisia, many testimonies and observations confirm this tendency and many nests were deposited even in sites not considered before by nesting activities.

In the frame of the project "Conservation of sea turtles in the Mediterranean region" coordinated by SPA/RAC and financially supported by MAVA (2018–2019), exploration of sandy beaches, along Tunisian coasts, looking for new and potential sea turtles nesting sites, was done.

Primarily results show that about 20 sites were identified as nesting sites for the loggerhead turtle, where previous or current nests were detected (**Figure 6**).

Moreover, the study of the quality of beaches patrolled indicates that the majority of them are favorable for nesting activity which allows us to consider them as potential nesting sites.

As noticed, surveys on Mediterranean coasts and especially on the coasts of the western basin show that nesting activity has increased recently from 1 year to another in many Mediterranean countries. Global warming phenomenon and increase of observation effort, mainly in the frame of science citizen, could be responsible for the extension of the nesting areas.

In Tunisia 65 nests were registered in 2020; 44 in Kuriat islands and 21 nests in other beaches of Tunisian coasts. **Figure 6** shows all nesting sites known in Tunisia for the moment.

Besides Kuriat islands and Chebba beaches, those considered nesting sites during 2016–2020 and where nests were really observed are presented in **Table 4**.

It is also noticeable that many other sites were discovered last 5 years in Tunisia.

### 4.4 First record of *Chelonia mydas* nesting in Tunisia

Following a testimony reported to NGB (association Notre Grand Bleu) on the discovery of a sea turtle nest on Rejich beach in summer of 2019, an expedition of experts and volunteers discovered that it is a nest of green turtle *Chelonia mydas*, hatchlings (**Figure 7**) gave more confirmation [11].

The nest was laid on August 03, 2019, on the beach of Rejich, 35.449871°N; 11.044676°E. Sea-nest distance was 19.5 m and cavity depth was 70 cm from the surface.

Sixty hatchlings reached the sea securely, while no hatchlings were found dead either inside or outside the nest. Hatching success was calculated as 54.5%

| Year | Governorate | Nesting activities |
|---|---|---|
| **Beach name: Zouaraa (Nefza)** | | |
| 2016 | Beja | *One nest deposited* [24] |
| 2018 | Beja | *One nest deposited* |
| 2019 | Beja | Two nests |
| 2020 | Beja | Four nests |
| **Beach name: Louka 2** | | |
| 2018 | Bizerta | Nesting activity detected |
| **Beach name: Fetha (Kelibia)** | | |
| 2018 | Nabeul | Flooded nest |
| 2020 | Nabeul | Nesting activity detected |
| **Beach name: Mansoura (Kelibia)** | | |
| 2018 | Nabeul | One nest |
| 2020 | Nabeul | One Lost nest |
| **Beach name: Petit Paris** | | |
| 2020 | Nabeul | Nesting activity detected |
| **Beach name: Dar Allouche** | | |
| 2020 | Nabeul | Nesting activity detected |
| **Beach name: Echraff** | | |
| 2020 | Mahdia | Two nests [41] |
| **Beach name: Chott Mariem** | | |
| 2020 | Sousse | One nest [41] |

**Table 4.**
*Nesting sites recorded and number of nest-laid during 2016–2020.*

**Figure 7.**
*Green turtle newborn hatching in Tunisia (Rejich Beach, 2019).*

(Eggshells/clutch size X 100). The remaining 50 eggs (unhatched eggs) were identified and included 14 early embryony stages (12.73%), 2 late stages (1.82%) and 34 unfertilized eggs (30.91%) [11].

### 4.5 Main threats on nesting beaches

- The nesting sites of small Kuriat and Chebba are highly frequented by swimmers during nesting season (**Figure 8**). The beaches are heavily used by humans and disturbance of the sand may have impeded the detection of turtle tracks or nests [2, 37].

- The black rat *Rattus rattus*, abundant on small Kuriat attacks hatchlings after emergence (**Figure 8**). Deratization undertaken by " Notre Grand Bleu " association in 2016 has resolved the problem.

**Figure 8.**
*Threats on nesting beaches: (1) Beaches of small Kuriat highly frequented (2) Hatchlings on the small Kuriat attacked by rat on their heads (3) Specimen of Ocypode cursor from the south of Tunsia 340] (4) Large deposits of seagrass (*Posidonia oceanica*) on the beaches of great Kuriat.*

- Sea gulls *Larus carchinans*, common on the Kuriat islands, seem to engender predation of hatchlings, mainly of those that emerged during daytime.

- The tufted ghost crab *Ocypode cursor* (**Figure 8**) is the only Ocypode species present in the Mediterranean Sea. The first observation of this crab, known to be a predator of sea turtles' hatchlings and eggs, in the Tunisian coasts was made by a hazardous observation of specimens emerging from their burrows in June 2018 near a nest of loggerhead turtle in the Kuriat Islands [42].

- The large deposits of phanerogam (*Posidonia oceanica*) on the beaches of Great Kuriat (**Figure 8**) mainly restrict the accessibility of nesting females to the site. These deposits of *Posidonia* hinder also the return of hatchlings to the sea after their emergence. However, the deposits constitute a natural protection of the beaches from waves and inundation.

- Light pollution concerns the two nesting beaches of Chebba [31, 32]. The light of the cornice and the port attract the hatchlings after their emergence. Hatchlings, disoriented, finish on the road behind the cornice where they are crushed by cars.

## 5. Bycatch of sea turtles

### 5.1 Bycatch and mortality rates

Bycatch of sea turtles in commercial fisheries occurs along Tunisian coasts but it is more important and more assessed in the Gulf of Gabès. This area is the most important fishing area that comprising about 50% of the Tunisian fishing fleet [48].

The high concentration of the fishing effort in the Gulf of Gabès has led to overexploitation of fish stocks and is contributing to bycatch of several threatened species as well as of many fish species. Besides, other pressures such as pollution and the spreading of alien species have contributed to the degradation of the ecosystems [48].

In this region, a large fishing fleet using many kinds of fishing gear operates during different seasons and targets a wide variety of commercially important species. These fisheries constitute the main threats to the sea turtles, through the direct mortalities and injuries associated with incidental bycatch. Although observations show that both species; the loggerhead turtle *Caretta caretta* and the leatherback turtle *Dermochelys coriacea* were captured by fishing gear in the Gulf of Gabès, the occurrence of the latter is very sporadic. Several studies have quantified the bycatch of sea turtles from various gears in the Gulf of Gabès (**Table 5**). The bycatch rates observed varied between the different gears and studies and were highest in gillnet fishery and pelagic longlines (**Table 5**). In terms of mortality, the highest rates were recorded by gillnet followed by bottom longlines (**Table 5**). The high mortality rates associated with gillnets, targeting sharks, it may be a result of the long soak time [48, 49]. In these fishing gear captured sea turtles might not be able to reach the surface to breathe and eventually die of asphyxia. In the Gulf of Gabès, sea turtles are bycaught because of the high degree of spatial overlap between the fishing grounds and the habitats of this species [43, 45, 49].

Although some studies exist on the impact of fishing gears on the *Caretta caretta* in the Gulf of Gabes region, the data remains fragmentary and the level of interaction of sea turtles with fisheries remains poorly assessed. Indeed, they are limited to

| Fishing gear | Observed catch rate | Estimated total captures | Recorded mortality | Reference |
|---|---|---|---|---|
| Pelagic longline | 0.823 (0.568–0.158) turtle/1000 hooks | 486 (335–683) | 0% | [46] |
| | 0.806 (0.802–0.810) turtle/1000 hooks | 437 (299–609) | 12.1% | [44] |
| | 0.25 turtle/1000 hooks | 100 | 3.44% | [45] |
| Bottom longline | 0.278 (0.179–0.415) turtle/1000 hooks | 733 (470–1090) | 33% | [43] |
| | 0.333 (0.236–0.591) turtle/1000 hooks | 142 (100–167) | 43.7% | [45, 47] |
| | 0.26 turtle/1000 hooks | 688 | 9.41% | [45] |
| Trawl | 0.0114 (0.0085–0.0143) turtle/ haul | 5458 ± 1652 | 3.3% | [52] |
| Gillnet | 0.527 (0.403–0.649)/ $km^2$/day | 444 (358–501) | 69.4% | [43] |
| | 0.63 (0.355–0.893)/ km/day | 3756 (1908–5902) | 92.06% | [49] |
| Trammelnets | 0.92 turtles $day^{-1}$ $km^{-1}$ | 2000 | | |

**Table 5.**
*Observed catch rates (95% C.I), estimated yearly captures (in numbers), and mortality rates of loggerhead turtles registered by different gears in the Gulf of Gabès.*

restricted areas in the Gulf of Gabès and concern only some metier. Furthermore, some discrepancies in the results regarding capture and mortality exist (**Table 5**). This could be due to the study period and the metier.

Recent Interviews conducted with fishermen in 19 ports along the Gulfs of Gabes and Hammamet revealed that sea turtle bycatch per unit effort (BPUE) was the highest for Gillnet (0.73 turtles/vessel/day), followed by pelagic longlines (0.6 turtles/vessel/day) (**Figure 9**). However, due to the trawlers' high fishing effort, the cumulative impact of the trawl nets was the highest with an estimated number of 11,740 (0–41,525.75) turtles caught per year in Tunisia [50]. These results corroborate the previous ones in that they show the same conclusions: Gillnets and pelagic longlines are the most gears impacting sea turtles.

Investigations made in the frame of the Medbycatch project (March 2019–January 2022) indicated that all fishing gears operating in the Gulf of Gabès affect sea turtles because of the overlap of their areas of activity with the feeding and wintering grounds (**Figure 10**). Overall, small-scale fisheries including nets and longlines are the main threats of sea turtles followed by trawlers.

The mortality at haul back of sea turtles is higher in small-scale fisheries (**Figure 11**). Among small-scale fisheries, gillnets have the highest mortalities (**Table 6, Figure 11**). However, fishermen are aware of the status of this species along the Tunisian coasts and all alive individuals at retrieval are generally released.

Considering the Tunisian coasts, MedBycatch project surveys showed that more than 70% of sea turtle catches occurred in the Gulf of Gabès followed by the Gulf of Hammamed (GSA13) where more than 27.5% of bycatches were reported. However, Sea turtle captures along the northern coasts (GSA, 12) were very scarce.

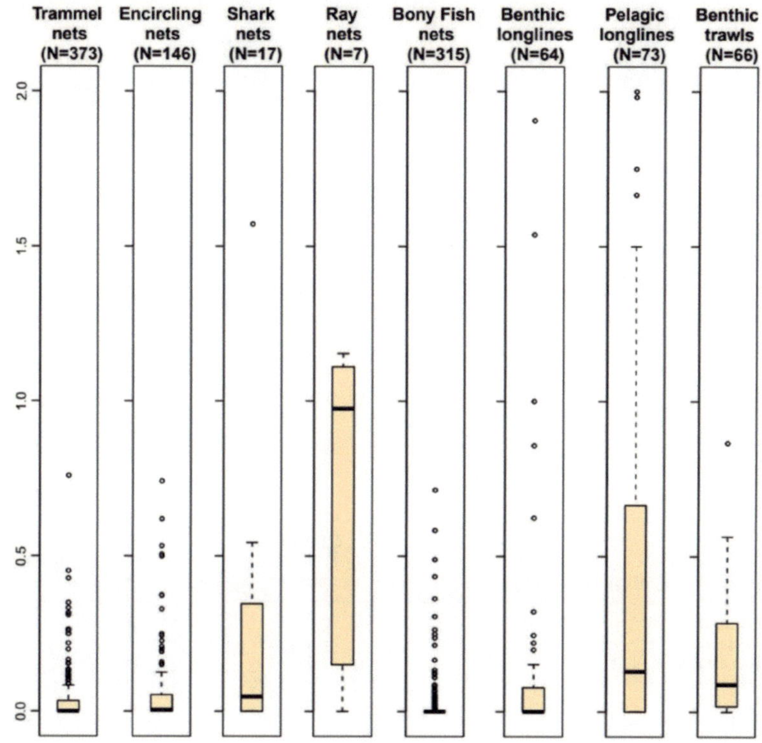


**Figure 9.**
*Sea turtle bycatch per day of fishing (BPUE) [50].*

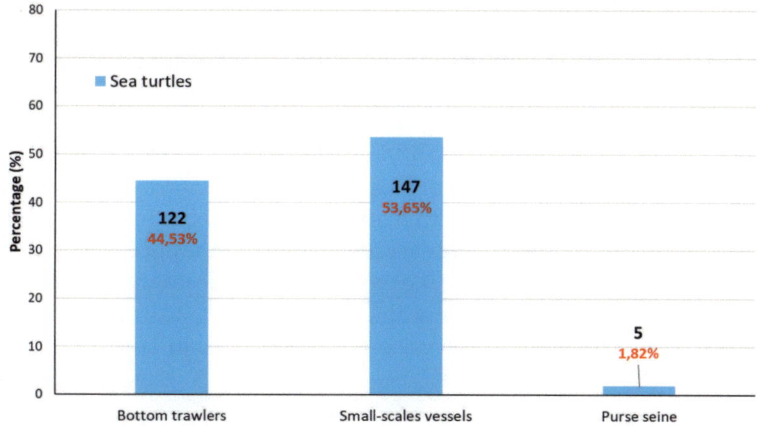


**Figure 10.**
*Percentage in number of individuals caught accidentally by gear type in GSA 14.*

## 5.2 Mitigation measures

In Tunisia, as in the rest of the Mediterranean basin, certain mitigation measures have been experimented with to reduce the capture and mortality of sea turtles. These measures include the use of circle hooks for longlines, soak time and depth. These studies remain preliminary and require further experiments for consolidation. On the

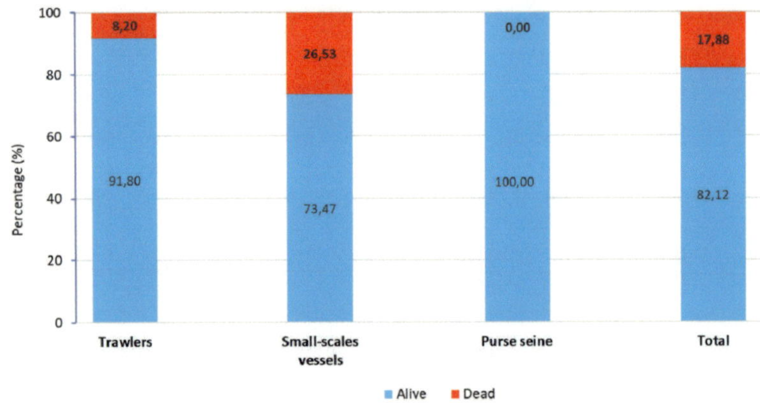


**Figure 11.**
*Mortality at haulback of* Caretta caretta *in percentage by gear type in GSA 14.*

| Species | GSA 12 | GSA 13 | GSA 14 | Total |
|---|---|---|---|---|
| *Caretta caretta* | 46 | 116 | 575 | 737 |
| *Chelonia mydas* | 0 | 0 | 2 | 2 |
| *Dermochelys coriacea* | 2 | 1 | 11 | 14 |
| Undetermined | 2 | 9 | 39 | 50 |

**Table 6.**
*Distribution of sea turtle strandings by GSA (2020–2022).*

other hand, no requirements to use these have been put in place, and the results are still controversial.

The use of circle hooks as a mitigation measure in pelagic longline fishery appear not promoting indeed it increases the catch rates of Sea turtles [45]. However, in the frame of Medbycatch project, reducing soak time to less than 12 hours for gillnets appears encouraging, in fact, the mortality rate decreases significantly when reducing the soak time. In the other hand, investigation showed that the bycatch rates of sea turtles decrease significantly with depth in gillnets, longlines and trawls suggesting that excluding fishery activities from coastal waters would reduce the impact of these fisheries on sea turtles. Fishing at depth beyond 30 m reduces largely the bycatch of sea turtles. Besides, increasing fishers and administration's awareness about sea turtles may help to improve the conservation of this species in Tunisia.

## 6. Sea turtle stranding

The study of stranded marine turtles was strengthened in the beginning of 2004 through the creation of the National Stranding Network (RNE). This program was included in the activities of the National Institute of Sciences and Technologies of the Sea (INSTM). Mainly three groups (researchers, veterinary doctors and students) have been set up to this effect, one based in the north, a second group based in the center and a third one in the south.

Many other actors are also involved in the records of stranding such as:

- The Sfax Faculty of Sciences (FSS);
- NGOs: The Notre Grand Bleu association, AJEM association, TunSea and many others;
- The World Wildlife Fund-North Africa.

For each stranding event, many data were registered, such as dates, Global Positioning System coordinates or location, kind of coast (sandy, rocky), body measurements, sex and species identification. Animals' conditions were reported as live animals, freshly dead, moderately decomposed (organs basically intact), advanced decomposition (organs not recognizable) and mummified/skeletal remains. Necropsy was performed on fresh and moderately decomposed animals. Tissues were taken for histopathology, toxicological and genetic analyses; they were frozen at −20°C or preserved in ethanol and stored at INSTM; the presence and nature of parasites and epibionts were noted.

In this chapter, we present the analyze of stranding data 2020–2022 (RNE commission report, 2023).

During this period, 806 stranding events were recorded including 740 (92%) loggerhead turtles *Caretta caretta*, 14 (1.73%) leatherback turtles *Dermochelys coriacea* and 2 (0.25%) green turtles *Chelonia mydas*. These results confirm that the loggerhead turtle is the most common species on the Tunisian coasts, the green turtle is, in fact, rare while the leatherback turtle is regularly observed in Tunisian waters [51]. 50 specimens have not been determined.

### 6.1 Sizes of stranded turtles

Among the 529 loggerhead turtles measured, the Curved Carapace Line of the smallest stranded is less than 20 cm, the largest was a female of 91 cm. The most represented size classes are those of juveniles to sub-mature (SCCL between 60 and 70 cm) (**Figure 12**).

This bell-shaped distribution (**Figure 12**) would be due to the fact that large juveniles are the most affected by accidental captures [43, 47], which would increase the number of stranding cases.

The low percentage of strandings of small turtles (SCCL <30 cm) would be due to the absence of interaction with fishing gear, especially during the first years of their life (Lost Years) when they seem to disappear. Before reappearing again 2 or 3 years later.

The low percentage of adults in the strandings seems to confirm the smallness of the nesting population in Tunisia.

For the 11 leatherback turtles *Dermochelys coriacea* measured, the Curved Carapace Line (CCL) of the smallest stranded is 138 cm, the largest was a female of 180 cm. The most represented sizes are between 140 and 160 cm) [52].

### 6.2 Sexes of stranded loggerheads

91.19% of the 227 adult turtles recorded are female

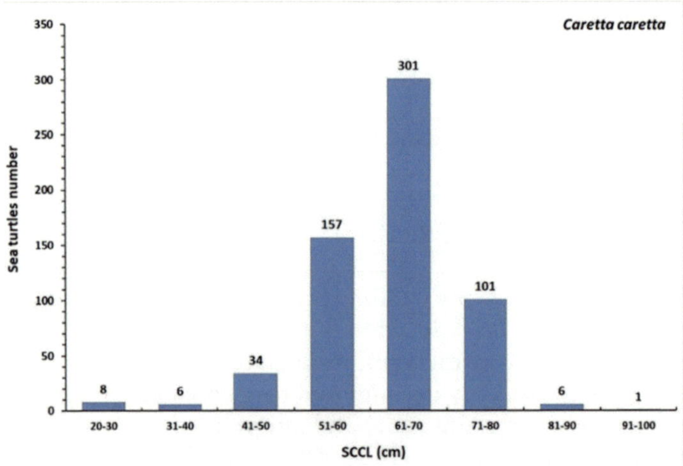


**Figure 12.**
*Number of stranded loggerheads by size classes 2020–2022.*

## 6.3 Spatio-temporal distribution of strandings

Strandings are more concentrated in southern Tunisia (GSA 14) (**Table 6**; **Figure 13**). This result confirms that:

- The Gulf of Gabès (GSA 14) is a foraging and wintering area for marine turtles in the Mediterranean [3, 6, 8]
- Turtle strandings are believed to be caused mainly by fishing. Indeed, the Gulf of Gabès area is an important maritime fishing area in Tunisia and is home to most of the country's fleet.

It should be noted, however, that the effort to prospect for strandings is not uniform for the three regions considered. It is more important in the Gulf of Gabès.

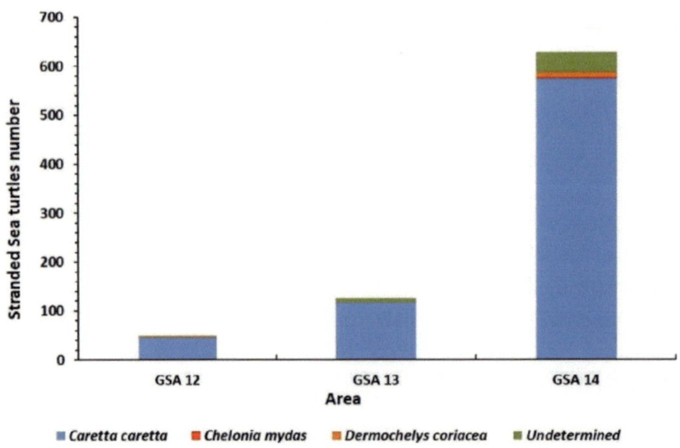


**Figure 13.**
*Spatial distribution of sea turtle strandings (2020–2022).*

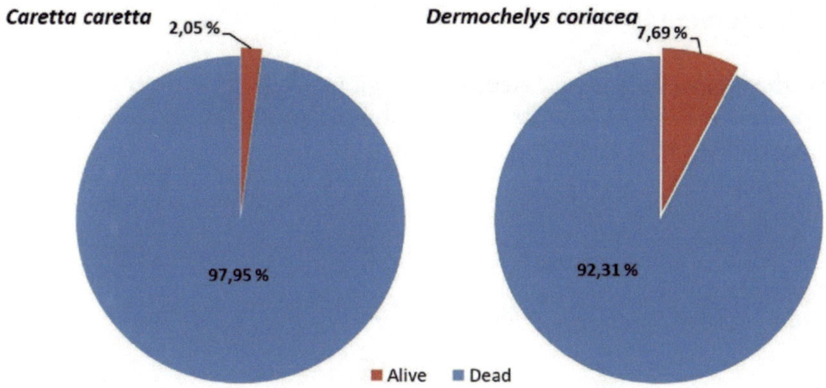


**Figure 14.**
*Distribution of strandings between dead and alive (2020–2022).*

Concerning the temporal distribution, strandings are more observed in spring and summer for the period 2020–2022 with a peak in May.

### 6.4 Causes of death

The majority of stranded turtles are dead, in fact, approximately 98% of loggerhead turtles and more than 92% of leatherback turtles arrive dead on the beach (**Figure 14**).

The interaction with fishing appears to be the major cause of marine turtle mortality when the mortality factor is known. Among the most impactful fishing gear, gillnets targeting elasmobranchs should be noted.

## 7. Conclusion

Sea turtles, once abundant in Tunisia at sea and on nesting beaches, were marketed for several uses. These uses are anchored in the collective memory of the population, especially islanders (food, shells for decoration and as cradles for babies, medicines, etc.). Faced with the unprecedented acceleration of the collapse of their global populations following several anthropogenic threats, they have become threatened with extinction and their conservation has become urgent. Bycatch seems to be very threatening. The available data on the assessment of bycatch clearly show that sea turtles are among the taxa most affected by accidental fishing on all Tunisian coasts and that several gears bring them back with different rates.

Several mitigation measures have been tested. We mainly cite the reduction in loggerhead catches in gillnets targeting elasmobranchs "Garracia" by implementing spatio-temporal solutions (depth, soak time). Similarly, catches of this species in trawls decrease with increasing depth. In addition, several conservation tools developed in Tunisia (legislation, rescue centers, stranding network, monitoring of nesting sites, etc.) have contributed to the protection of sea turtles in Tunisia, and even in the Mediterranean Sea. Fishers automatically release turtles caught accidentally at sea and even intervene in the release of turtles entangled in nets and in informing the authorities and care centers.

## Acknowledgements

The authors gratefully acknowledge the Regional Activity Center for Specially Protected Areas (SPA/RAC) for its contribution in funding our research's on vulnerable species and the cost of editing this chapter.

## Author details

Mohamed Nejmeddine Bradai[1*], Imed Jribi[2], Bechir Saidi[3] and Samira Enajjar[4]

1 Faculty of Sciences of Sfax, ASCOB-Syrtis, Sfax, Tunisia

2 Faculty of Sciences of Sfax, University of Sfax, Sfax, Tunisia

3 Faculty of Science and Technology of Sidi Bouzid, University Campus Agricultural City, Sidi Bouzid, Tunisia

4 National Institute of Sciences and Technologies of the Sea (Sfax Center), Sfax, Tunisia

*Address all correspondence to: mednejmeddine.bradai@instm.rnrt.tn

## References

[1] APAL. Vers une Gestion Intégrée du Littoral Tunisien. Tunisie: Ministère de l'Environnement et du Développement Durable Agence de Protection et d'Aménagement du Littoral; 2015. 47 p

[2] Bradai MN, Jribi I. Tunisia. In: Casale P, Margaritoulis D, editors. Sea Turtles in the Mediterranean: Distribution, Threats and Conservation Priorities. Gland, Switzerland: IUCN Press; 2010; 294 p

[3] Argano R. Preliminary report on western Mediterranean Sea turtles. Annual report on WWF project 1474. Rome; 1979. 19 p

[4] Bradai MN. Les captures accidentelles de Caretta caretta au chalut benthique dans le golfe de Gabès. Rapport Commission Internationale Mer Méditerranée. 1992;**33**(1):285-285

[5] Bradai MN, Bentivegna F, Jribi I, El Ouaer A, Maatoug K, El Abed A. Monitoring of loggerhead sea turtle *Caretta caretta*, in the Central Mediterranean via satellite telemetry. In: Demetropoulos A, Turkozan O, editors. Proceedings of the Second Mediterranean Conference on Marine Turtles (Barcelona Convention-Bern Convention-Bonn Convention), Kemer, Turkey; 2005. 2009. pp. 54-57

[6] Gerosa G, Casale P. Interaction of Ma-rine Turtles with Fisheries in the Mediter-ranean. Regional Activity Centre For Specially Protected Areas. Boulevard de l'Environnement, BP 337 - 1080 Tunis Cedex – Tunisie. 1999. 59p

[7] Laurent L, Nouira S, Jeudy De Grissac A, Bradai MN. Les tortues marines de Tunisie: Premières données. Bulletin of the Herpetological Society of France. 1990;**53**:1-17

[8] Laurent L, Lescure J. L'hivernage des tortues caouannes *Caretta caretta* dans le Sud tunisien. Revue d'Ecologie (Terre vie). 1994;**49**:63-86

[9] Margaritoulis D. Post-nesting movement of loggerhead sea turtles tagged in Greece. Rapport Commission Internationale Mer Méditerranée. 1988;**31**(2):284

[10] Margaritoulis D, Argano R, Baran I, Bentivegna F, Bradai MN, Caminas JA, et al. Loggerhead turtles in the Mediterranean sea: Present knowledge and conservation perspectives. In: Bolten AB, Witherington BE, editors. Loggerhead Sea Turtles. Washington DC: Smithsonian Books; 2003. pp. 175-198

[11] Ben Ismail M, Jribi I, Kaska Y, Ben Nakhla L, Amir Ben Fradj A, Dibej M, et al. The westernmost green turtle (*Chelonia mydas*). Nest Recorded in the Mediterranean from Tunisia M.T.B. 2022;**1**:19-22

[12] Rees AF, Saad A, Jony M. Discovery of a regionally important green turtle *Chelonia mydas* rookery in Syria. Oryx. 2008;**42**:456-459

[13] Saad A. Importance of Lattakia Beach (Syria) as nesting area for marine turtles: Results of seven years of field survey. Scholarly Journal of Agricultural Science. 2012;**2**:108-110

[14] Saad A, Soulaiman A, Alkusairy H. Marine turtle nesting survey and stranding assessment from Tartus to Syria's border with Lebanon. Natura Croatica. 2020;**29**(Suppl. 1):51-57

[15] Monconduit P. Situation de Ia pêche maritime en Tunisie au 1er janvier

1927. Bulletin Station Océanographique Sa-lammbô. 1927;**6**

[16] Blanc M. Faune tunisienne, dactylographiée en trois parties : Mammifères, Oiseaux, Reptiles et Batraciens. In: Bulletin mensuel de la Société linnéenne de Lyon. 1935;**9**:48-149

[17] Servonet J. Les pêches dans le Golfe de Gabès. Revue maritime et coloniale. 1889;**101**:142-161

[18] André L. Les îles Kerkennah. In: Tome IBLA, editor. Etude d'ethnographie tunisienne et de géographie humaine. 1961. 405 p

[19] Marinkelle CJ. Drie Skinkssoorten van Tunesië. Lacerta. 1959;**17**:64-66

[20] SPA/RAC - ONU Environnement/ PAM. In: Par Bradai MN, Jribi I, editors. Stratégie nationale pour réduire le commerce illégal des tortues marines en Tunisie. Tunis: SPA/RAC, Projet MAVA Tortue Marine; 2020. 79 p

[21] Ellouze G. Contribution à l'étude de la reproduction de la tortue caouanne *Caretta caretta* Linnaeus, 1758. DEA Université de Tunis II; 1996; 111 p

[22] Knoepffler LP. Une curieuse anomalie de la carapace chez *Caretta caretta* L. Vie et Milieu. 1962;**13**(2):237-331

[23] Parent GH. Quelques observations écologiques sur l'herpétofaune de l'île de Djerba. Naturel belg. 1981;**62**:122-150

[24] Bradai MN, Karaa S. Première mention de la nidification de la tortue caouanne *Caretta caretta* sur la plage Zouaraa (Nord de la Tunisie). Bulletin de l'Institut National des Sciences et Technologies de la Mer de Salammbô. 2017;**44**:203-206

[25] Bradai MN. La nidification de la tortue marine *Caretta caretta* dans le Sud-Est de la Tunisie. Rapport Commission Internationale Mer Méditerranée. 1995;**34**:237-237

[26] Bradai MN. Les tortues marines en Tunisie. Etat de connaissances et recommandation de conservation [thesis]. Tunisia: University of Sfax, Faculty of Sciences of Sfax; 2000; 47 p

[27] Jribi I, Bradai MN, Bouain A. Marine turtles nesting in Kuriat islands (Tunisia) in 2000. Marine Turtle Newsletter. 2002;**96**:4-6

[28] Jribi I, Bradai MN, Bouain A. The loggerhead turtle nesting activity in Kuriat islands (Tunisia) in 2001. Bulletin de la Société Herpétologique de France. 2002;**102**:43-47

[29] Jribi I, Bradai MN, Bouain A. The loggerhead turtle nesting activity in Kuriat islands (Tunisia): Assessment of nine years monitoring. Marine Turtle Newsletter. 2006;**112**:12-13

[30] Jribi I, Bradai MN. Sex ratio estimations of loggerhead sea turtle hatchlings at Kuriat islands, Tunisia. Can minor nesting sites contribute to compensate globally female biased sex ratio? The Scientific World Journal. 2014;**2014**:8

[31] Ben Hassine J, Escoriza D. *Caretta caretta* in Tunisia: Natural history and report of a new regular nesting area. Herpetological Review. 2013;**44**(4):557-561

[32] Jribi I. Loggerhead turtle *Caretta caretta* nesting activity in Chebba (Centre Tunisia): Assessment, problems and recommendations. Indian Journal of Geo-Marine Sciences. 2017; V 46 (1). pp. 163-169.

[33] SPA/RAC - ONU Environnement/ PAM. In: Par Jribi I, Bradai MN,

editors. Plan d'Action National pour la Conservation des Tortues Marines. Tunisie: SPA/RAC, Projet MAVA Tortue Marine; 2020. 36 p

[34] Hrizi M. Contribution à la recherche des sites de nidification des tortues marines en Tunisie. Mastère de recherche en Sciences du Vivant. Tunisie: Faculté des Sciences de Sfax-Université de Sfax; 2019. 66 p

[35] Benabdi M, Belmahi AE. First record of loggerhead turtle (*Caretta caretta*) nesting in the Algerian coast (southwestern Mediterranean). Journal of the Black Sea/Mediterranean Environment. 2020;**26**(1):100-105

[36] Demetropoulos A, Hadji-chrostophorou M. Manuel on Marine Turtle Conservationin the Mediterranean. UNEP (MAP) SPA/IUCN/CWS/Fisheries Department, MANRE (Cyprus); 1995; 63 p, 24 plates

[37] Bradai MN, Jribi I, Karaa S. Tunisia. In: Casale P, Hochscheid S, Kaska Y, Panagopoulou A, editors. Sea Turtles in the Mediterranean Region: MTSG Annual Regional Report 2020. Report of the IUCN-SSC Marine Turtle Specialist Group; Switzerland. 2020:286-302

[38] Chaieb O, Elouaer A, Maffucci F, Bradai MN, Bentivegna F, Said K, et al. Genetic survey of loggerhead turtle *Caretta caretta* nesting population in Tunisia. Marine Biodiversity Records. 2010;**V 3**:1-6

[39] Girondot M, Fretey J. Mise au point d'une fiche de description de sites de ponte. Rapport de fin de contrat rédigé à la demande du Ministère de l'environnement; 1996

[40] Hochscheid S, Maffucci F, Abella E, Bradaï MN, Camedda A, et al. Nesting range expansion of loggerhead turtles in the Mediterranean: Phenology, spatial distribution, and conservation implications. Global Ecology and Conservation. 2022;**2022**(38):1-14

[41] Chaieb O, Maatouk K, Dhraief MN, Bradai MN. First evidence of Loggerhead sea turtle (*Caretta caretta*) nesting events on the Tunisian beaches Echraff and Chott Mariem (south Central Mediterranean). Herpetological Review. 2022;**53**(1):6-9

[42] Karaa S, Jrijer J, Bradai MN, Jribi I. New record of *Ocypode cursor* (Linnaeus, 1758) (Crustacea: Decapoda: Ocypodidae) from the Tunisian coasts, the Central Mediterranean Sea. Journal of the Black Sea/Mediterranean Environment. 2019;**25**(1):101-107

[43] Louhichi M, Girard A, Jribi I. High loggerhead (*Caretta caretta*) bycatch rate along with several endangered target species: Two reasons to look for alternative to traditional large-mesh bottom-set gillnets (Garrasia) for more sustainable fisheries in the Gulf of Gabès (Tunisia). Sustainability. 2024;**16**:3713

[44] Bradai MN, Saidi B, Enajjar S, Karaa S. Pêcheries aux palangres de fond et de surface dans le Golfe de Gabès. Rapport final. MoU ACCOBAMS N°07/2016/LB6410; 2017. 55 p

[45] Jribi I, Bradai MN, Bouain A. Incidental captures of sea turtles by longline in the Gulf of Gabès (South Tunisia): Comparative study between bottom longline and surface longline. Scientia Marina. 2008;**72**(2):337-342

[46] Echwikhi K, Jribi I, Bradai MN, Bouain A. Effect of type of bait on pelagic longline loggerhead interaction in the Gulf of Gabès south of Tunisia. Aquatic Conservation:

Marine and Freshwater Ecosystems. 2010;**20**(5):525-530

[47] Echwikhi K, Jribi I, Bradai M.N, Bouain A. Interactions of loggerhead turtle with bottom longline fishery in the Gulf of Gabès, Tunisia. Journal of the Marine Biological Association of the United Kingdom. 2012;**92**(4):853-858

[48] Béjaoui B, Ben Ismail S, Othmani A, Ben Abdallah-Ben Hadj Hamida O, Chevalier C et al. Synthesis review of the Gulf of Gabes (eastern Mediterranean Sea, Tunisia): Morphological, climatic, physical oceanographic, biogeochemical and fisheries features. Estuarine, Coastal and Shelf Science. 2019:395-408

[49] Echwikhi K, Jribi I, Bradai MN, Bouain A. Gillnet fishery -loggerhead turtle interactions in the Gulf of Gabes, Tunisia. Journal of Herpetology. 2010;**20**:25-30

[50] Louhichi M, Girard A, Jribi I. Fishermen interviews: A cost-effective tool for evaluating the impact of fisheries on vulnerable sea turtles in Tunisia and identifying levers of mitigation. Animals. 2023;**13**:1535

[51] Bradaï MN, El Abed A. Présence de la Tortue Luth Dermochelys coriacea dans les eaux Tunisiennes. Rapport Commission Internationale Mer Méditerranée. 1998;**35**

[52] Jribi I, Bradai MN, Bouain A. Impact of trawl fishery on marine turtles in the Gulf of Gabès, Tunisia. Journal of Herpetology. 2007;**17**:110-114

Chapter 3

# Vulnerable Marine Vertebrates along the Egyptian Mediterranean Coast: Challenges to the Anthropogenic, Fisheries Bycatch and Climatic Changes

*Mahmoud M.S. Farrag, Ahmed A. Mustafa, Ismail M. Abdelazim and Yassein A. Osman*

**Abstract**

This chapter introduces the status of the marine vertebrate's vulnerability in the Mediterranean Sea, Egypt, since their vulnerability has increased mostly due to fisheries activities, climatic changes besides, anthropogenic pollution as macroplastics and extra.... Cartilaginous fishes suffer negatively from fishing activities as bycatch mostly by trawl, longliners, and purse seiners. Small-scale fisheries affect the nursery ground capture small-sized fish. The sunfish Mola mola has exposed to vulnerability due to fishing activity. Sea turtles are also included in this issue due to the fishing activity particularly setting nets, trading, and consumption of turtles still encourages the fishers to keep them for selling. Thus, the absence of awareness among fishers has negatively affected their conservation, particularly loggerhead and green turtles. Urbanization led to a reduction in nesting, while macroplastic directly affected adults due to bags and bottles and entangling in the remains of fishing nets. Climatic changes affected the ecological and food web movement influencing sea turtles and marine mammals, causing frequent beached and sighted mammal's cases such as fine whales. The topography fluctuates from deep to poor shallow water as the Delta region in Egypt, participates in stranding rates together with shipping activities. The mitigation and conservation measures are recommended (powerful legislations and the awareness of public societies and fishers, good handling of discarded bycatch to ensure their healthy status is good after release to the Sea, fishing tools modification as longlines, time reduction of setting nets). Pollution and macro-plastic materials shall be decreased, and the remains shall be removed.

**Keywords:** marine vertebrates' vulnerability, fisheries bycatch, elasmobranchs, sea turtles, mammals, conservation measures, Mediterranean Sea, Egypt

## 1. Introduction

The Mediterranean Sea biodiversity is suffering from severe alterations because of anthropogenic pressure and its interaction with the effects of climate change [1, 2]. Climate changes have increased globally, becoming obvious in all countries; such changes can directly or indirectly affect biodiversity in both flourishing animals and even those being sighted occasionally. It has also affected the migration and transfer of species from one place to another, and large animals such as mammals can experience stranding together due to shipping and anthropogenic activities [3]. In the Mediterranean Sea, fishing is severely affecting ecosystems [4, 5] and is a major threat to large vertebrates of conservation concern such as sharks [6], sea turtles [7], and cetaceans [8].

From the other point of view, the large marine vertebrates exposed to vulnerability are large fishes including large bony ones which are characterized by low populations and cartilaginous species (i.e., sharks and rays). Their characteristics of low fecundity, late maturity, and slow growth make them more likely to be endangered due to any impacts [9, 10]. The Cartilaginous species are the most endangered in the Mediterranean Sea. Among 73 assessed species in the Mediterranean, the Red List status of Chondrichthyans shows that 39 (53% of all) are critically endangered or vulnerable [11]. Overfishing, non-selective fishing practices, and habitat degradation are leading to dramatic declines of these species in the Mediterranean Sea. According to Ferretti *et al.* [6], the sharks in the Mediterranean Sea have declined by more than 97 percent in number and 'catch weight' over the last 200 years. This decline is greatly affecting food webs throughout this region and could hold serious implications for the entire marine ecosystem. The overexploitation of such megafauna is mostly due to bycatch exploitation [12, 13].

According to Dulvy et al. [10], a decline in elasmobranch populations has been observed throughout the world and was particularly marked in the Indo-Pacific and Mediterranean Seas. The loss of some shark and batoid populations from aquatic ecosystems has socioeconomic and ecological consequences [14, 15]. The Mediterranean elasmobranch species are mainly coastal and benthic (80%), which makes them vulnerable to fishing activities concentrated mainly on the coasts [16]. The small-scale fisheries in the Mediterranean Sea, considered a source of mortality for early-life stages of elasmobranch species, are mostly in nursery areas so that they can be recognized among the greatest threats to nontargeted bycatch megafauna [17, 18]. The decline of elasmobranchs was detected due to intensive fishing activities in the Mediterranean basin throughout the coastal and pelagic waters from the bast as in the Gulf of Lions [19], the Adriatic Sea [20], and the Tyrrhenian Sea [21]. Hence, most of the elasmobranchs in the Mediterranean Sea had been captured as bycatch [22].

In Egypt, small-sized specimens of large rays were detected from the coastal zone off Alexandria by small-scale fisheries bycatch (gill nets), which included five species: *Dasyatis pastinaca*, *D. tortonesei*, *Rhinoptera marginata*, *Glaucostegus cemiculus* [23]. In Egypt, small-sized specimens of large rays were detected from the coastal zone of Alexandria by small-scale fisheries bycatch (gill nets), which included five species: *Dasyatis pastinaca*, *Dasyatis tortonesei*, *Rhinoptera marginata,* and *Glaucostegus cemiculus* [23]. Moreover, several studies have mentioned sharks and rays as bycatch from shallow and deep-water trawlers [3, 24–26]. Sharks and rays were investigated from the bycatch of various fishing gears in abundance for the common species [27]. The other large vertebrate fish, *Mola* has a low population and slow movements that make it easier to be included in the bycatch of various fishing gears. In Egypt, it was observed at different times and because of its big size, it has been acceptable for trading. This species is also listed on the global Red List of Vulnerable Species by the International Union for Conservation of Nature (IUCN). The reports of such fish species are mentioned in the present study.

The other large marine vertebrates exposed to decline and vulnerability in the Mediterranean Sea due to various factors are sea turtles. The situation of marine turtles differs from that of others species, such as elasmobranchs and mammals, due to the probability that life and sea turtles can be saved—even if they are caught in the bycatch. In fact, trading and eating consumption demand is among attractive reasons to increase its fisheries impact. They represented nutritional, economic, and spiritual resources to many human communities at least 7000 years ago [28]. From the seventeenth century onwards, marine turtle consumption increased because of factors such as the increase in the human population, increased fishing, and capture capacity, and more efficient trade that responded to wider market demand [29–32]. Six of the seven extant marine turtle species are categorized as threatened on the IUCN Red List [33]. The consumption and trade of marine turtles was banned or regulated through several international conventions and national legislation [17, 34, 35]. Nevertheless, the consumption of turtle meat still occurs, legally or not, in many places [32, 36]. They are declining due to factors such as fisheries impacts [30].

The two common species in the Mediterranean are the loggerhead turtle *Caretta* and the green turtle *Chelonia mydas.* They are from bycatch impact, where the latter has a relatively small population restricted to the easternmost Mediterranean [37]. Both loggerhead and green turtles have been sold in several fish markets along the Mediterranean coast—that is, Alexandria, Abu Qir, Burullus, Port Said, and Damietta—since at least the beginning of the twentieth century [38, 39]. This trading has increased the impact of *via* encouraging some fishers to avoid sea turtles' again to water. The governmental efforts and restrictions have been recently applied for the places of Sea turtles trading, which affected positively to be mostly focused on Alexandria, followed by Abu Qir, instead of the various places mentioned above.

Egypt is a signatory to several conventions for nature conservation, and several national regulations that cover marine turtles were issued in the 1980s and 1990s. The most important is Environmental Law 4 of 1994, which states that 'killing, capturing, transportation, selling, nest destruction, and display of an endangered species either dead or alive is prohibited when Egypt is a signatory to an International Convention and national regulations between 1980 and 1990'. This bycatch and the illegal turtle trade probably affect several marine turtle populations in the Mediterranean. However, despite this regulation, egg consumption was observed in 1998 [40], and trade in marine turtle meat in Alexandria fish markets went on at least until 1999, with 18–25 turtles slaughtered per week in the Anfoushi fish market [38, 41]. The government increased the enforcement of laws in the Alexandria fish markets in 1999 [42]. On the other hand, the conservation of marine turtles is extremely crucial in the Mediterranean Sea, while Egypt is considered the requested spot in the eastern part based on studies by Broderick *et al.* [43] and Rees et al. [44]. They stated that the nesting of green turtle females has been tracked by satellite from Cyprus and Syria to Egypt, as the Egyptian waters are considered foraging grounds and migratory corridors for green turtles, while Cyprus and Syria are considered two of the important three nesting areas in the Mediterranean. The genetic markers indicate that loggerhead turtles found in Egypt also originate from Turkey and, possibly, Greece [30], which represent most of the major nesting grounds for this species [45].

Egypt is highly affected by the fisheries impact of different fishing gears directly by its fishing mostly as bycatch and through the remains of fishing gears, which are considered sometimes as pollution and involved in plastic influencing. The high effect of fishing gear was observed for the setting nets, which entangled the Sea turtles and submerged them in water for more than 2 to 3 days until they died with no chance of being released [22, 39]. The data were not from direct observation on board; they were mostly taken from interviews with fishermen and traders. The following survey of impacts on Sea turtles reported the impacted Sea turtles onboard and other factors affecting sea turtles had been expanded over trading and bycatch including pollution, particularly macroplastics, technology, and urbanization along the coast which negatively affected the nesting places to be reduced particularly in western Alexandria and reduced in a number of green turtles, while the leatherback turtle became very rare. The cases and causes were reported in this study, which focuses on the Egyptian Coast of the Mediterranean Sea.

Cetaceans comprise an important topic among vulnerable species. Several cases of stranding were observed, particularly in the eastern part of the Mediterranean Sea during the past few decades, possibly due to the occurrence of climate changes combined with increasing anthropogenic activities [3, 46]. Few cetaceans such as fin whales and sperm whales were found over a wide range across the Mediterranean Sea [3, 47–49] and combined with their vulnerability to anthropogenic pressures [50–52]. The IUCN, in collaboration with the Agreement on the Conservation of Cetaceans of the Black Sea, Mediterranean Sea and Contiguous Atlantic Area (ACCOBAMS), has listed 21 species and three subspecies of cetaceans in the Mediterranean Sea and Black Sea [33].

Marine mammals' vulnerability is an important issue of concern for biodiversity, and numerous cases of stranding have been documented in the Mediterranean Sea with respect to the eastern part [53–55]. Monitoring cetacean populations is important and is a priority for conservation worldwide. They are impacted by various anthropogenic pressures [47, 49, 50]. Knowledge of eight cetacean species was cited in the eastern part of the Mediterranean Sea including stranding and live samples from Israel [56], Greece [53], and Turkey [54]. Recent studies in Egypt were cited by [3, 46]. Several marine mammal species have then been observed along the Egyptian Coast of the Mediterranean Sea, either in life or stranded specimens.

Among the vulnerable cetaceans in the area are six species in the family Delphinidae: the bottlenose dolphin *Tursiops truncatus* (Montagu, 1821), which has an extended distribution on the eastern coast, with no difference in abundance on either side of the Mediterranean [53]; the striped dolphin (*Stenella coeruleoalba* (Meyen, 1833)); Risso's dolphin (*Grampus griseus* (G. Cuvier, 1812)); the long-finned pilot whale (*Globicephala melas* (Traill, 1809)), which is the largest dolphin species in the Mediterranean and is usually limited in its distribution to the western basin; the short-beaked common dolphin (*Delphinus delphis* Linnaeus, 1758); the rough-toothed dolphin (*Steno bredanensis*), and Cuvier's beaked whales (*Ziphius cavirostris G. Cuvier*, (1823)). Some other species were considered visitors, such as killer whales (*Orcinus orca* (Linnaeus, 1758)), which sometimes move from the Atlantic to the Mediterranean Sea to feed on tuna species. Other species found included the dwarf sperm whale, which is extremely rare in the Mediterranean basin [33]. These species are protected in the Mediterranean through different agreements and conventions such as ACCOBAMS for the conservation of the Cetaceans of the Black Sea, Mediterranean Sea, and Continuous Atlantic Area. The objective of this chapter is to highlight the vulnerability of large marine vertebrates—that is, elasmobranchs and large bony fishes and sea turtles and cetaceans—with respect to fisheries impacts and bycatch as well as the general negative factors and the challenges toward conservation and mitigation measures in the Mediterranean Sea, Egypt.

## 2. Materials and methods

### 2.1 Study area

The area conducted for the investigation in this chapter is the Egyptian Coast of the Mediterranean Sea. The investigation was based on the survey of bycatch impacts and threats to large marine vertebrates, which extended from Al-Arish eastward (34°12′36″E and 31°15 ′00″N) to Al-Sallum westward (**Figure 1**). The coast is extended to about 950 km long where the coastline can be divided into three zones. The eastern zone, about 180 km long, extends from Port Said to Rafah and is mostly characterized by rocky bottoms with sandy features in front of Al-Arish. It has a continental shelf area of about 1500 $km^2$. The central zone, about 300 km long, extends from Alexandria to Port Said with a muddy bottom as the majority of the delta formation beside the sandy and some rocky zones are scattered. The western zone is about 450 km long. Extending from Alexandria to Al-Sallum, this zone is characterized by extremely deep slopes and high irregular form [57].

### 2.2 Data collections

The data in the present chapter were collected from the published and unpublished works together with a questionnaire with the scientific community and fishers.

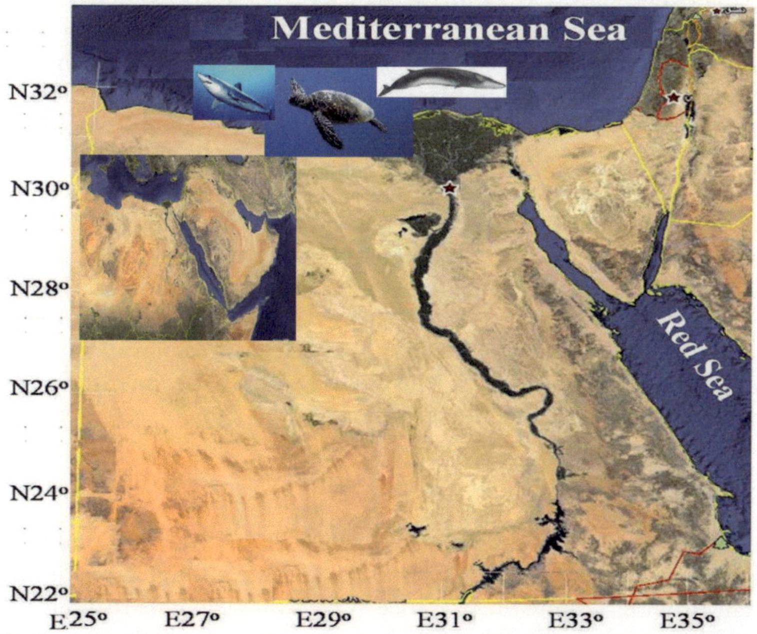


**Figure 1.**
*General map of the Egyptian Coast of the Mediterranean Sea, Egypt.*

Published data were obtained mostly from the reported data on the Egyptian Coast of the Mediterranean Sea for elasmobranchs have been mentioned as bycatch [3, 23, 24, 26]; for the Sea turtles the historical view of the status and impacts on, where obtained based on [40]; for marine mammals particularly the marine cetaceans, the data of the threatened and stranded mammals were obtained from those reported [3, 46, 58, 59]. The recorded cases and unpublished were identified according to the above-mentioned references, adding some citations [33, 60–64].

The factors of the impacts and threats were summarized according to published data mentioned above, the current survey and questionnaire with fishers and public people and the scientific community and stakeholders. The recommendations and challenges to decline the bycatch impacts and threats on the large marine vertebrates were summarized based on the international efforts reported by scientists such as Bradai et al. [22] and authorities (IUCN, [33]; Food and Agriculture Organization (FAO) [65]; UNEP-SPA/RAC [66]; GFCM [67]; and SPA/RAC [68]), and the local efforts by scientist [3, 27]; authorities as Environmental Affairs, Ministry of the Environment, Egypt and LERPDA Agency, Egypt, finally current vision of the authors obtained from the causes of impact and the questionnaire with all stakeholders.

## 3. Results (threatened marine vertebrates)

### 3.1 Fishes

#### *3.1.1 Bony fishes*

The large marine vertebrates are exposed to vulnerability and endangered including some bony fishes, mostly the sunfish, which live in the oceanic water with slow movements that make it easy to be caught. Its weight can reach 1500 kg, which is acceptable for fishers and traders. Due to its low number, it became a threatened species. It has a high sense of macroplastics during feeding while the major factor is fishing activity as bycatch. It was reported several times along the coast from different areas. It was caught before 2010 with no confirmed data from the Egyptian Mediterranean water from Rashid, then caught also from the Rashid area by a purse seiner in 2013 and landed in the Alexandria Landing Centre (**Figure 2**). It was then caught from Rashid again in 2016 (900 kg) and has been transformed into NIOF. It was caught again from Port Said in January 2019 (400 Kg), followed by a specimen caught from Burullus in March 2019 (800 kg). A small specimen of sunfish was caught recently in June 2024 from the Egyptian Mediterranean water and landed in the Alexandria Landing Centre (45 kg) (**Figure 3**). These specimens were collected as a bycatch of bottom trawlers and purse seiners.

#### *3.1.2 Cartilaginous fishes*

The cartilaginous/elasmobranchs including sharks and rays are among vulnerable fishes which are caught as bycatch. The bycatch species suffered from decline due to the impact of fisheries despite the presence of laws to prevent its fishing and trading. A lot of species are included in the IUCN Red List, most of which are large sharks and rays. The sharks and rays face sometimes further challenges in the Mediterranean basin [69, 70] due to fishing pressure. At least 53% of the Chondrichthyes species

**Figure 2.**
*The sunfish caught from the Egyptian Mediterranean water from Rashid landed in the Alexandria Landing Centre in January 2013 (Mahmoud Farrag).*

**Figure 3.**
*The sunfish caught from the Egyptian Mediterranean water in June 2024 and landed in Alexandria Landing Centre (45 kg) (Mahmoud Farrag).*

found in the Mediterranean are classified by the IUCN as 'vulnerable' and 'endangered' [11, 71], and a large proportion of species (20%) are still classified as 'data deficient'.

In the Mediterranean and Black Seas, the Food and Agriculture Organization (FAO) reported 49 shark species belonging to 17 families and 5 orders—that is, Hexanchiformes, Squaliformes, Squatiniformes, Lamniformes, and Carcharhiniforms [72]. According to the IUCN Red List's Shark Specialist Group, 24% of shark, skate, and ray species are considered threatened species facing extinction, with the urgent need for conservation efforts [11]. The increasing consumption of shark products, along with the shark's fishing vulnerabilities, led to a decrease in certain shark populations [73]. Most reasons for the impact are the fisheries, such as bycatch of different fishing gears, particularly those that catch small-sized specimens in the coastal areas where these areas are mostly considered nursery ground, caught by small-scale fisheries. **Table 1** and **Figure 4** show that most species are vulnerable to fisheries demand as bycatch, as mentioned in Farrag and Adel [27]. From this table, some of the species were not listed in the red list of IUCN due to some species of sharks still present in quantities in Egypt, while others were included as endangered and vulnerable as listed by IUCN. Generally, the Egyptian sector, as the southeastern part, needs more studies and attention, unlike the central and western parts of the Mediterranean Sea. This may be due to the topography of delta and insufficient fishing operations in deeper water.

### 3.2 Sea turtles

Sea turtles in the Mediterranean Sea are categorized as required to be conserved. They suffer from trading, eating consumption, and fishing activities as bycatch. Two species of sea turtles, the loggerhead turtle *Caretta* and the green turtle *Chelonia mydas*, are common, whereas the leatherback sea turtle is considered a rare species. In addition to fishing, most of this impact is attributed to trading and acceptability for food consumption in different regions along the Egyptian coast, particularly the

| **Family** | **species** | **Common Names/(IUCN status)** |
|---|---|---|
| Carcharhinidae | *Carcharhinus altimus* (Springer, 1950) | Big nose shark/(Data deficient) |
| Carcharhinidae | *Carcharhinus plumbeus* (Nardo, 1827) | Sandbar shark/(Endangered) |
| Carcharhinidae | *Carcharhinus falciformis* (Bibron, 1839) | Silky shark |
| Glaucostegidae | *Glaucostegus cemiculus* (Geoffroy St. Hilaire, 1817) | Blackchin guitarfish/ (Endangered) |
| Hexanchidae | *Hexanchus griseus* (Bonnaterre, 1788) | Blunt nose six gill shark/ (Least concern) |
| Lamnidae | *Isurus oxyrinchus* (Rafinesque, 1810) | Shortfin mako/(Critically Endangered) |
| Scyliorhinidae | *Scyliorhinus canicula* (Linnaeus, 1758) | Small spotted cat shark/(Least concern) |
| Sphyrnidae | *Sphyrna lewini* (Griffith and Smith, 1834) | Scalloped hammer head shark/(Not applicable) |
| Triakidae | *Galeorhinus galeus* (Linnaeus, 1758) | Tope shark/(VulnerableA2bd) |
| Triakidae | *Mustelus asterias* Cloquet, 1821 | Starry Smoothhound/(VulnerableA2bd) |
| Triakidae | *Mustelus* (Linnaeus, 1758) | Common smoothhound/(Vulnerable) |
| Triakidae | *Mustelus punctulatus* Risso, 1827 | Blackspotted Smoothhound/(Vulnerable) |
| Dasyatidae | *Dasyatis pastinaca* (Linnaeus, 1758) | Common stingray/(Vulnerable) |
| Mobulidae | *Mobula mobular* (Bonnaterre, 1788) | Spintail devil ray/(Endangered) |
| Myliobatidae | *Aetomylaeus bovinus (Geoffroy St. Hilaire, 1817)* | Bull ray/(Critically Endangered) |
| Rhinopteridae | *Rhinoptera marginata* (Geoffroy-Saint-Hilair, 1817) | Lusitanian cownose ray/(Least concern) |

**Table 1.**
*The common sharks and rays exploited intensively along the Egyptian Coast of the Mediterranean Sea including vulnerable, endangered, and critically endangered species according to the IUCN list of the Mediterranean Sea, 2016 [11].*

regions of Abu Qir and Alexandria, with the absence of awareness. This was reported previously in somewhat similarity with Nada and Cásale [39]. The survey and questionnaire reflected that the problem is not only for sea turtles captured by fishers for trading but also the fishers themself eat the sea turtles onboard as hidden data (based on personal communication and observation of the remains), then throwback the sea turtles remains. Moreover, the recent effect of climatic changes and urbanization has negatively affected sea turtles and their nesting places.

Nada and Cásale [39] introduced a report in 2011 on sea turtles based on a survey conducted with a high number of fishermen in different regions along the coast

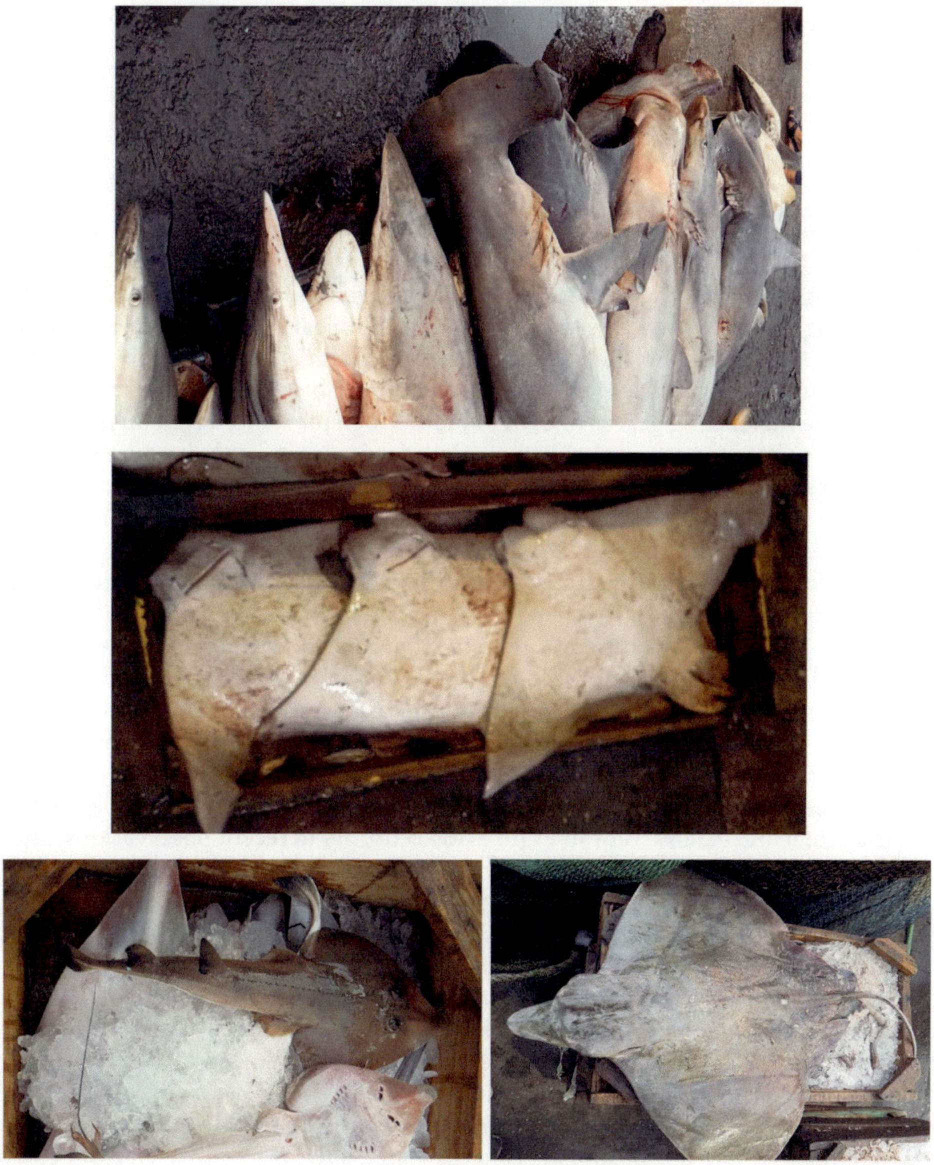

**Figure 4.**
*Various shark species from the Alexandria Landing Centre, guitar shark, and large rays which are vulnerable along the Egyptian coast of the Mediterranean Sea. (Mahmoud Farrag).*

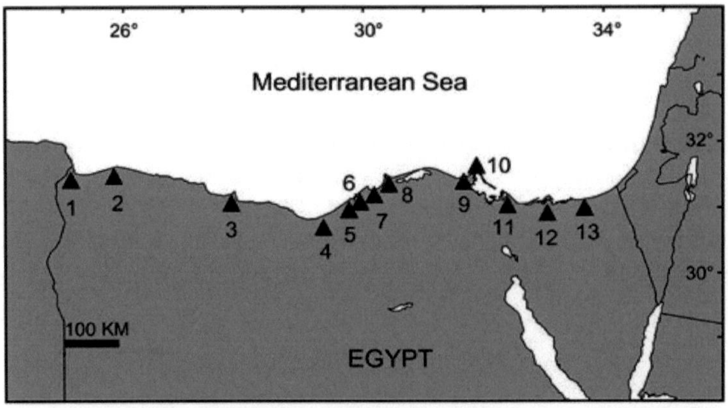


**Figure 5.**
*The Mediterranean coast of Egypt including 13 places where surveys were carried out. Western Region: 1, Al-Sallum; 2, Sidi Barrani; 3, Marsa Matruh; 4, Al-Hammam; 5, Alexandria Governorate (Al-Max, Anfoushi, Abu Qir); Central or Delta Region: 6, Al-Maadiya; 7, Edco; 8, Rosetta; 9, Izbat al-Burj; 10, Damietta; Eastern or Sinai Region: 11, Port Said; 12, Lake Bardawil; 13, Arish. [39].*

(**Figure 5**). As they stated, there is a reduction in nesting areas in the eastern region, while nesting is rare in the western region and negligible in the central one. Among the catch rates declared by fishermen and the official statistics on the fishing fleet, over 5000 turtles per year are estimated to be captured by trawlers, longliners, and set netters. Since this fishing gear is known to induce high mortality rates, an important proportion of these turtles, probably many hundreds, can die only because of capture. As stated by Mohamed Nada, Head of the Save the Sea Turtle Project, sea turtle populations have declined throughout the Mediterranean, particularly off the coast between Alexandria and al-Sallum. Various reasons include direct exploitation, the destruction of nesting sites due to the rapidly expanding tourist industry, and a lack of awareness among the public.

According to the most recent and current situation, the trading of turtles continues despite its fishing being prohibited by the Egyptian lows by the Ministry of Environment and General Authority for Fish Resources and Development. The reduction of observed species may be due to these restrictions; however, tourism, urbanization, and anthropogenic activities have affected negatively the nesting areas to be reduced (**Figure 6A, B**). **Figure 6A** shows wide green circles in the past; however, they turn into yellow color in **Figure 6B** according to recent observations and due to urbanization and anthropogenic activities. From 2013 until now, many cases have been observed of nesting area changes and adults in the fish markets. The most frequently observed species is the loggerhead sea turtles.

The present survey illustrates that urbanization has increased not only from Alexandria to Al-Sallum but also extends to the eastern part of the coast toward Port Said. Trading of sea turtles still operates in illegal form, particularly in Alexandrian regions where people like to eat its meat. Even though, loggerhead turtles are abundant, green turtles are preferred. The effect of exploitation on sea turtles was observed in both nesting areas and on small and adults *via* fisheries bycatch. However, it extends to its hot spot areas in the water, such as two areas in front of al-Max, Montazah, and Marassi. These are extensive areas with sea turtles. Most of these areas are resting areas located in shallow areas after deep water far away from the coast. Based on the survey, most fishing activities negatively affect sea turtles,

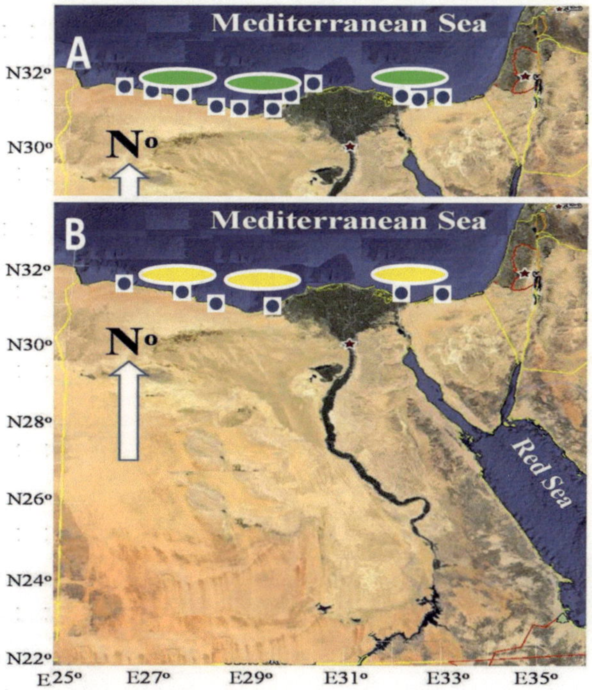


**Figure 6.**
*Major areas of sea turtles occurrences (small circles) were in wide green circles previously (A) and reduced areas in yellow circles (B) recent years.*

including longlines, trawlers, and purse seiners, which are mostly set nets that stay more than 1 day in the water. Pollution, particularly macroplastic pollution, and the remains of fishing nets float on the water's surface or on the bottom. The present survey illustrated that the intensive observation of sea turtles along the Egyptian coast was conducted from December to May in various sizes along the coast considering the hot spot areas mentioned previously. During the season, a various activities were observed including positive efforts to conserve the sea turtles and negative impacts *via* traders and fishers in different areas from Alexandria to Port said, while the major area for trading and consumption extended from Alexandria to Abu Qir encouraging the transportation of sea turtles illegally.

### *3.2.1 The efforts applied to conserve sea turtles*

Despite the impact of various factors on sea turtles, various efforts have been applied related to the laws of the Ministry of Environment, where Egypt mitigates the trading and fishing of sea turtles even as bycatch. The General Authority for Fish Resources and Development (GAFRD) during 2012 has issued decree No. 151 to prevent catching sea turtles; however, the strong monitoring increased by 2017, leading to an increase in awareness among some fishers and traders, particularly in Damietta and Port Said together with positive efforts of environmental managers in Ashtom El Gamil protectorate to rescue several sea turtles and return it to the sea. More efforts have been exerted by other society associations such as the team of Turtle and Wildlife Rescue in Alexandria, supervised by Ms. Mai Hamada, Head of the team,

| Species (No) | Area | Date | Observer | Status of impact/ Action |
|---|---|---|---|---|
| Green turtle (2) | Alexandria, | Feb. 2017 | Fish market | Bycatch (Dead) |
| Green turtle (1) | Alexandria | 2017 | Rescue team | Returned to the Sea water |
| Green turtle (1) | Alexandria (Fish market) | 2018 | Rescue team | Returned to the Sea water |
| Loggerhead (1) | Alexandria | 2018 | Rescue team | Bycatch (Eaten) |
| Loggerhead (1) | Sidi Abd Al Rahman | Oct. 2018 | Fisher, trawl at night | Bycatch (Returned to Sea) |
| Loggerhead (15) | Sidi-Krir to Al-Alameen | Nov. 2018 | Fishers, trawl (5 days) | (Fish market, Alexandria, Eaten) |
| Green turtle (1 small) | Abu-Qir | 2019 | Rescue team | Nesting area (Returned to Sea) |
| Loggerhead (1 small) | Abu-Qir | 2019 | Rescue team | Nesting area (Returned to Sea) |
| Loggerhead (1) | Alexandria | March, 2019 | Rescue team | Returned to the Sea water |
| Green turtles (11) | Alexandria | Dec. 2019 | Fish market | Four dead and seven shells |
| Green turtles (1) | Port Said | 2021 | Rescue team | Bycatch (Returned to Sea) |
| Logger head (small) | Alexandria | Feb. 2022 | Fish market | Bycatch (Returned to Sea) |
| Green turtle (1) | Alexandria | April, 2023 | Fishers | Still life |
| Green turtle (1) | Alexandria | May, 2023 | Rescue team | Bycatch (Returned to Sea) |
| Green turtle (1) | Alexandria | June, 2023 | Rescue team | Bycatch (Returned to Sea) |
| Loggerhead (1) | Alexandria | June, 2023 | Rescue team | Bycatch (Returned to Sea) |
| Loggerhead (1) | Sidi krir | Aug. 2023 | Fishers, trawl | Bycatch (Returned to Sea) |
| Green turtle (1) | Alexandria | Aug. 2023 | Rescue team on beach | Returned to the Sea water |
| Loggerhead (8) | Al-Arish, | Aug. 2023 | Hotel beside beach | Small individual Returned to Sea |
| Loggerhead (1) | Alexandria | Sept. 2023 | On beach | Dead impacted by boat) |
| Loggerhead (1) | Alexandria | Sept. 2023 | Fish market | Bycatch (Returned to water), Team |
| Green turtle (1) | Alexandria | Sept. 2023 | Fish market | Bycatch (Returned to Sea), Team |
| Green turtle (1) | Port Said | Sept. 2023 | Fishers, fish market | Bycatch (Returned to water by environmental managers) |
| Loggerhead (1) | Alameen | Nov. 2023 | Fisher, trawlers | Bycatch (Returned to Sea) |
| Logger head (20) | Alexandria | Feb. 2024 | Fisher, Fish market | Bycatch (Eaten) |
| Green turtle (15) | Alexandria | Feb. 2024 | Fisher, Fish market | Bycatch (Eaten) |
| Loggerhead (1) | Rashid | March, 2024 | Purse Seine | Bycatch (Returned to Sea) |
| Green turtle (1) | Alexandria | April, 2024 | Fish market | Bycatch (Eaten) |
| Loggerhead (1) | Alexandria | April, 2024 | Fish market | Bycatch (Eaten) |
| Loggerhead (1) | Rashid | May, 2024 | Purse Seine | Bycatch (Returned to Sea) |

| Species (No) | Area | Date | Observer | Status of impact/ Action |
|---|---|---|---|---|
| Green turtle (2) | Alexandria | May, 2024 | Fishers in water | Still life |
| Leather back (1) | Alexandria | May, 2024 | Fishers in water | Still life |
| Green turtle (2) | Alexandria | July, 2024 | Fish market | Bycatch (Eaten) |
| Loggerhead (2) | Alexandria | July, 2024 | Fish market | Bycatch (Eaten) |
| Green turtle (24) | Alexandria | Dec. 2024 | Fish market | Bycatch (Eaten) |

**Table 2.**
*The observed cases of sea turtles from 2017 up to 2024 have had various impacts along the Egyptian Mediterranean coast (Mahmoud Farrag).*

rescued several sea turtles many times, which are considered a positive point in conservation. However, the fisheries community requires more awareness of these species without hiding data far away from public monitoring. The scientific authorities, such as the National Institute of Oceanography and Fisheries (NIOF), Alexandria, have participated, rescued, and released several sea turtles to the seawater again. Many cases of vulnerability rescued by fishermen directly from onboard, along with other cases of sea turtle observation, are shown in **Table 2** and **Figure 7**. Based on this data, a high number of sea turtles were caught as bycatch directly by longliners, bottom trawl, and purse seiners while setting nets and macroplastics remains/marine litters caused damage and death for individuals. Other causes of sea turtles' vulnerabilities are pollution and urbanization which decline the nesting places and rest for adults.

(IUCN [33]; Green turtle and Loggerhead turtle are endangered; Leatherback recognized as critically endangered).

### 3.3 Threatened cetaceans

Marine mammals, with respect to cetaceans in the Mediterranean Sea have faced vulnerability due to fishing operations as a major activity, as well as shipping and navigation. In the eastern part of the Sea particularly the southeastern such as Egypt, few cases of whales were observed in the past. However, in recent decades, a lot of cases have appeared, both alive and dead. A recent study of stranding and vulnerability of marine mammals was conducted [3, 46]. They reported some cases of vulnerable marine mammals on the Egyptian Coast of the Mediterranean Sea from 2013 until 2018. The cases were identified and described; they were identified as Fin Whale (*Balaenoptera physalus* Linnaeus, 1758), the sperm whale (*Physeter macrocephalus*), Cuvier's beaked whale *Ziphius cavirostris*, the common bottlenose dolphin (*Tursiops truncates*), and the rough-toothed dolphin (*Steno bredanensis*) which was stranded on the Gamasa coast and was not completely confirmed. The sixth species was the California sea lion (*Zalophus californianus* Lesson, 1828) which was reported for the first time and can be escaped from the ship during transportation.

The most frequent occurrence of vulnerable mammals is the common stranded Fin Whale *B. physalus*, which was reported in 1936 along the Egyptian coast of Rosetta (Rashid) 1936, and a sample was deposited at the museum of the National Institute of Oceanography and Fisheries (NIOF) at Alexandria. After a long time, more than 70 years, this species was observed again in different regions during a short period, as shown (**Table 3**; **Figures 8A,B** and **9**). It was observed in 2008 on

**Figure 7.**
*Some cases of observed vulnerable Sea turtles in Egypt Mediterranean Sea rescued due to efforts and awareness, and others showing negative plastics and marine litter.*

the Baltim coast, on the Alamein coast in 2014, and in the Alexandria marina in July 2016, where it was found in a shallow small area for 1 day. It was then reported on the Alexandria coast in 2018 and in the Ghalion zone, Metubes, Kafr al-Sheikh, as a beached whale in 2021 [3, 46].

The repeated sightings in the eastern Mediterranean coast of Egypt in recent years were higher than those in previous time and can be related also to the increase in observation efforts, which can indicate that this area has become a visiting area for feeding or breeding or through pathways of migration together with the probability of

| Species | Common name | Location | Area | Date | Status | Reference |
|---|---|---|---|---|---|---|
| *Balaenoptera physalus* (Linnaeus, 1758) F: Balaenopteridae | Fin Whale | 31°481′187″N, 30°379′576″E | Rosetta (Rashid), Mediterranean Sea, Egypt | 1936 | Beached and stranded | Farrag et al. [3] |
| | | 31°598 ′653″N, 31°149 ′858″E | Baltim, Medit. Sea, Egypt | 2008, 2014 | Beached and stranded | Farrag *et al.* [3] |
| | | 30°842 ′515″N, 28°956 ′279″E | ElAlamein, Medit. Sea, Egypt | 2014, 2015 | Beached and stranded | Farrag *et al.* [3] |
| | | 31°289 47″N, 30°021909″E | Alexandria (Montazah), Medit., Sea, Egypt | June 2016 | live | Farrag *et al.* [3] |
| | | 31°234 ′522″N, 29°949 ′516″E | Azur beach, Alexandria, Medit., Sea, Egypt | January 2018 | Beached and stranded | Farrag *et al.* [3] |
| | | 30° 25 ′585″N, 31°905 ′27″E | Ghalion zone, Metobus, Kafr El-Sheik | Februry 2021 | Beached and stranded | Farrag et al. [46] |
| *Physeter microcephalus*Linnaeus, 1758F: F; Physeteridae | Sperm Whale | 31°15 ′20.8″N, 27°15.4 ′59″E | Cleopatra beach, Matrouh Province, Medit., Sea, Egypt | May 2016 | Beached and stranded | Farrag et al. ([3] |
| *Mesoplodon europaeus* (Gervais, 1855) F: Ziphiidae | Gervais' Beaked Whale | 31°39′36″N, 27°30′36″E | north coast, Matrouh Province, Medit., Sea, Egypt. | September 2018 | Beached and stranded | Farrag *et al.* [3] |
| *Tursiops truncates* (Montagu, 1821) F: Delphinidae | Common Bottlenose Dolphin | 31°241 ′461″N, 29°929 ′438″E | Gelim, Medit., Sea, Egypt | May 2017. June 2017. September 2018 | Beached and stranded | Farrag et al. [3] |
| | | 31°210 ′511″N, 29°911 ′586″E | Elshatby Medit., Sea, Egypt | November 2017 | Beached and stranded | Farrag *et al.* [3] |
| | | 31°18′0″N, 30°10′0″E | Abu Qir | Deember 2018 | Beached and stranded | Farrag *et al.* [3] |
| | | | | Every year up to 2024 | | |
| *Steno bredanensis* (G. Cuvier in Lesson, 1828) Family: Delphinidae | Rough-toothed Dolphin | 31°541 ′229″N, 31°523 ′487″E | Gamas, Medit., Sea, Egypt | June 2018 | Beached and stranded (Not confirmed) | Farrag et al. [3] |

| Species | Common name | Location | Area | Date | Status | Reference |
|---|---|---|---|---|---|---|
| *Monachus* (Hermann, 1779) Family: Phocidae | Monk Seal | 31°087 ′258″N, 28°027 ′378″E | Ras Elhekma, Medit., Sea, Egypt | 1990 | live | Farrag et al. [3] |
| | | 31°15 ′20.8″N, 27°15.4 ′59″E | Matrouh, Medit., Sea, Egypt | 1990 | live | Farrag et al. [3] |
| | | 31°33′09″N 25°09′27″E | Sallum, Medit., Sea, Egypt | 2017 | live | Farrag et al. [3] |
| *Zalophus californianus* | Sea lion | 31°525 ′531″N, 31°826 ′797″E | Damietta, Medit., Sea, Egypt | 2013 | live | Farrag et al. [3] |

**Table 3.**
*The sights and strandings of marine mammals along the Egyptian coast of the Mediterranean Sea.*

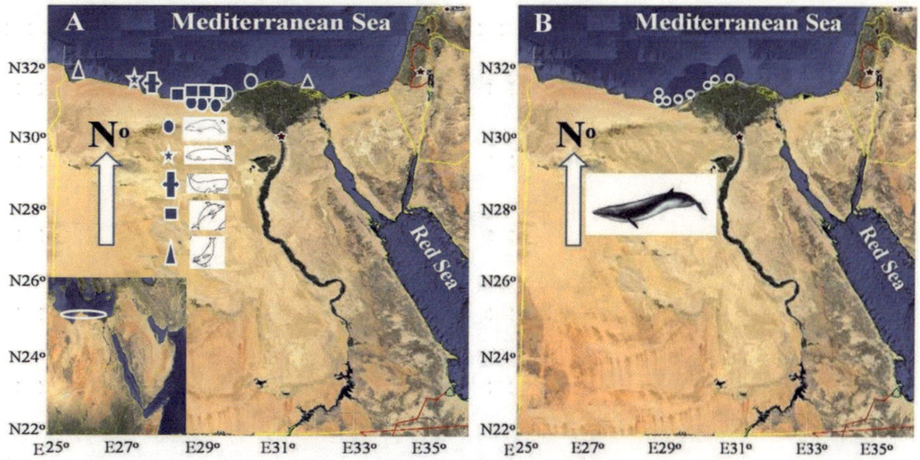


**Figure 8.**
*A: The sightings of the marine mammals on the Egyptian coast of Egypt [3], B; show the repeated occurrence of the fine whale [46].*

climate change effect. However, its appearance and stranding indicate its vulnerability due to different reasons: death caused by hunger because of oligotrophic conditions in the eastern part than in the western part or collision with ships and the topography of the Nile Delta region, which has become shallower throughout the huge region of the Egyptian coast. This species is particularly abundant in the Corso-Ligurian basin and the Gulf of Lion. However, the population declined from 900 individuals in 1992 to 147 in 2009 in the Pelagos Sanctuary, which is the largest marine protected area for Mediterranean marine mammals [33]. This indicates that this species has become a vulnerable and endangered species in agreement with IUCN [33], and this whale may fall victim to ship strikes, particularly from high-speed ferries [33].

The other frequent stranded whale in recent decades is the sperm whale, *Physeter Macrocephalus* Linnaeus, 1758. This whale was observed as a stranded one in May 2016 on one of the western Egyptian coast beaches in Matruh Province (31°15′20.8″N, 29°15.4′59″E) (**Figure 10**) [3]. It was then reported by the Egyptian Ministry of Environment in June 2023 near the Port Said coast as a dead whale. The population of this species in the Mediterranean was at least 30 individuals until the 1950s. According to the IUCN [33] sperm whales are mainly threatened by entanglement in fishing gear, particularly swordfish drift gillnets and tuna driftnets, ship strikes, and disturbance from maritime traffic. They mostly occur on continental slopes where they can dive extremely deep, more than 1000 m to feed, mostly on cephalopods. According to Goffman *et al.* [74], some sperm whale strandings have been cases of accidental straying or drifting of floating carcasses. In August 2023, other sperm whales were stranded on the coast of Lebanon. The stranded cases of sperm whales seem to be in summer, which can indicate its migration from west to east and facing some struggles during their journey such as changes in topography.

The third most frequent stranded whale is Cuvier's beaked whale (*Ziphius cavirostris*), which was misidentified as Gervais' beaked whale (*Mesoplodon europaeus*) in 2018 by Farrag et al. [3] due to some hidden features in the pictures related to the remarkable features for correct identification and actual conservation along the Egyptian Mediterranean coast in association with anthropogenic and climate changes. The correction of identification was done by Farrag et al. [46] (**Figure 11**). According

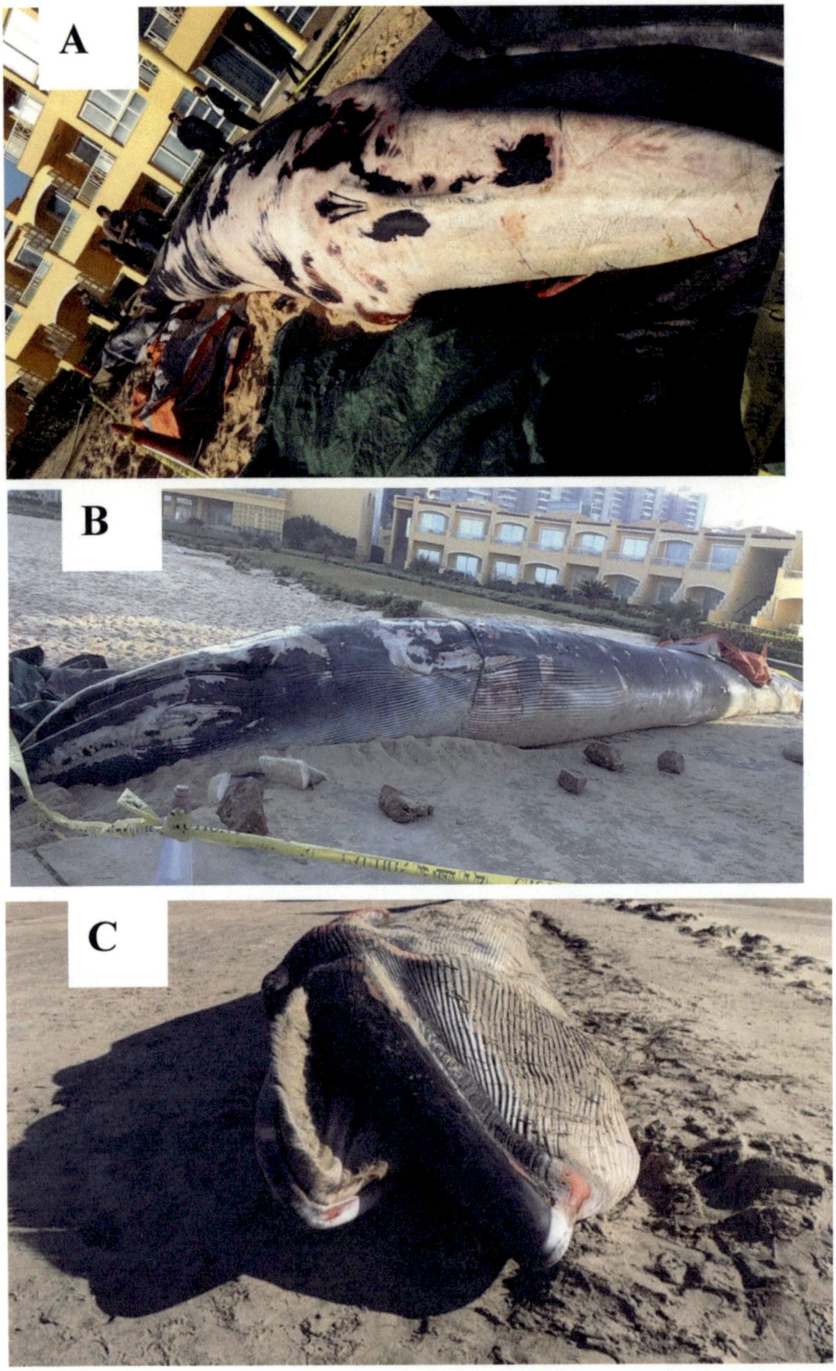


**Figure 9.**
*Fin Whale (B. Physalus (Linnaeus, 1758)) stranded on different zones along the coast of Egypt, Mediterranean Sea. (A, B and C in Farrag et al. [3, 46]).*

to Podestà et al. [75], the presence of two beaked whales (genus *Mesoplodon*) that were stranded alive along the coast of the French Riviera indicated the presence of *Mesoplodon* in the Mediterranean. They have documented 238 Cuvier's beaked whale

**Figure 10.**
*The stranded Sperm Whale (Physeter Macrocephalus Linnaeus, 1758) in [3].*

strandings in the Mediterranean Sea consisting of 327 animals in the Mediterranean Sea over a 204-year period. The stranding locations generally occur where regions of steep bathymetry occur close to the coast [75]. The current stranding location is in the western part of Egypt along the Mediterranean coast, and it is characterized by relatively deep water and the presence of steep bathymetry close to the coast. Moreover, the absence of this species in different regions of the Mediterranean basin may indicate that this species is not endemic and may be a visitor as stated by the IUCN [33]. In June 2024, another stranded Cuvier's beaked whale (*Z. cavirostris*) was observed nearly in the same area in the western part of the Mediterranean Sea, Egypt. It was observed dead on the north beach, Egypt, with a bad status.

The fourth frequently stranded marine mammal was a common bottlenose dolphin (*Tursiops Truncatus* (Montagu, 1821)). A specimen was found on the coast of al-Shatby, Alexandria (31°210′511″N, 29°911′586″E) in June 2017 (**Figure 12**). It was indicated that its death was due to a collision with a boat because of the presence of bruises on the front of the body [3]. This accident was previously reported in May 2017 on the Gleem Beach in Alexandria City near the previously mentioned stranding

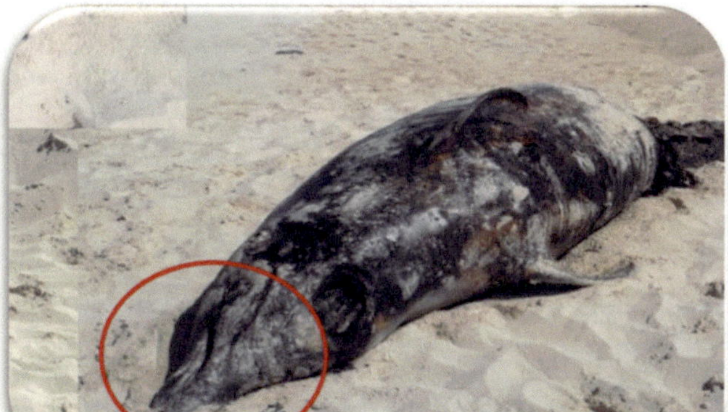

**Figure 11.**
*The stranded Gervais' beaked Whale Ziphius cavirostris from the Mediterranean coast, Egypt [3, 46].*

**Figure 12.**
*Stranded common bottlenose dolphin T. truncates from the Mediterranean coast, Egypt [3].*

site, but at that time, the dolphin was found in a stressed state with no caudal fin, which can have been served by a boat engine. A more recent incident occurred including the same species on 22 December 2018 at the Abu Qir beach (31°18′0″N, 30°10′0″E). The following data were taken from this male specimen: SL 205 cm, TW 300 kg, and 44 teeth in each jaw; it is a male specimen [3]. It has been noted that some predators attacked such dolphins and removed the lower part of the body; this could have been the reason for their deaths. The bottlenose dolphin is ranked as the most frequently vulnerable species; every year 1 or 2 individuals are observed in standing until 2024. From the other view, a lot of fishermen, particularly trawlers and purse seiners, when they find the dolphin in their catch, they release it again into the water.

This species is distributed in tropical and temperate waters around the world, including coastal and offshore environments, bays, estuaries, and lakes [74]. It is found in the Red Sea [75] and is a well-known dolphin in the Mediterranean basin [33, 76, 77]. It occurs in the waters of Albania, Algeria, Croatia, Cyprus, France, Gibraltar (United Kingdom), Greece, Israel, Italy, Montenegro, Morocco, Slovenia, Spain, Tunisia, and Turkey [77]. Many of the Mediterranean areas inhabited by these dolphins are subject to intensive human use, and this species is included on the Red List of Vulnerable Marine Mammals [33]. The occurrence of stranding indicates the tendency of some vulnerable marine mammals to become endangered.

For other mammalian stranded species, the rough-toothed dolphin (*Steno bredanensis*) (G. Cuvier in Lesson, 1828)), and even a California sea lion (*Zalophus californianus* Lesson, 1828), an exotic species likely escaped from a ship during exportation and transportation. Moreover, the *humpback whale* (*Megaptera novaeangliae*) (Borowski, 1781), was observed on 19 February 2021, according to the Ministry of Environment, this whale was a female with a 12-meter-long found dead on a beach in the Mediterranean Sea in Kafr al-Sheikh, Northern Egypt. Despite the unclear features of the humpback whale, there were doubts that it could have occurred, particularly after the increase of migration and climatic changes that affect the biodiversity that started from the small species attracting large animals. It is worth mentioning that all stranded whales in both frequent observation and single or few occurrences are listed on the IUCN Red List.

## 4. Conclusion

The current chapter has concluded that the Egyptian Mediterranean coast is a hot spot area for various factors that negatively affect biodiversity with respect to large marine vertebrates, and it needs real mitigation measures, conservation criteria, and scientific attention. The conclusion here is divided into two portions including the main factors of threats and vulnerability, while the second portion is the recommendations and challenges.

### 4.1 Factors of the vulnerability influencing on large marine vertebrates in the Mediterranean Sea, Egypt

- In the Mediterranean Sea, fishing is a major threat to large marine vertebrates and is affecting negatively on the ecosystems.
- Trading and consumption are among the major fisheries bycatch impacts for shark, rays, and sea turtles.
- Absence of awareness among fishers toward some large bony fishes such as *Mola*. The most reported specimens of *M. mola* were moved to the fish landing and sold for consumption. Few specimens were moved to the authorities as NIOF after the death just for preservation, in spite of the fishers having the chance to rescue it.
- Urbanization along the coast has affected negatively on the ecosystem, particularly the nesting and resting areas of sea turtles and other components of the ecosystems.
- Macro plastic pollution including bags, remains of the fishing gears, mostly gillnet, longlines, and other marine litters are affecting negatively on the Sea turtle, and cartilaginous fishes and may extend to small cetaceans.
- Fishing gear such as set nets and trawlers, mostly deep-water trawlers affects negatively on sharks and rays where the nets still in water for a longtime, and the catch includes dead and weak specimens. The dead cases of entangled specimens for longtime underwater may extend to small cetaceans.
- The operation of small-scale fisheries near the shore causes the mortality for early-life stages of elasmobranchs during the nursery period.
- For large cetaceans as whales, Egypt started to be a spot of vulnerability and stranding due to climatic changes that have affected the movement of passways due to changes in the ecological parameters and food web relative to whales.

- The shipping and navigation road may be affected by large whales, such as sperm whales, which are usually stranded in western Alexandria. At this part, the continental shelf is near the shore, moving from deeper to shallow. Fin whales have been stranded in different localities in the delta region as it is a shallow and muddy area that makes it disturbs and loses its road during migration as well as the area suffers from the decline of food matter.

- Poor knowledge of biological information and fishery databases, lead to the limits of the assessment and management plan of elasmobranchs.

### 4.2 Recommendation and mitigation challenges

- The recommendations and challenges that could be used to mitigate the vulnerability and bycatch impacts on large marine vertebrates in Egypt are summarized based on the subregional cases (published and unpublished) with modifications as follows.

- More attention and restrictions are required to mitigate the bycatch influence on large marine vertebrates. While the regulations to prohibit elasmobranchs and sea turtles have been released by the Ministry of Environment and Agency LFRPDA, in Egypt. However, these need to follow up where the impact is still via catching and trading.

- The awareness of public society toward nature conservation is urgently needed. The public society can reduce the consumption and trading of these animals, particularly sea turtles, which are mostly concentrated and still in Alexandria illegally. The awareness and monitoring should be paid to traders, fishers, and consumers.

- The intensive time of Sea turtles in Egypt and Mediterranean Sea starting from December until May should be considered to prevent its catching and trading.

- Direct observation, scientific data, and questionnaires with fishermen, traders, and coastal guards for beached specimens and nesting places of Sea turtles are required, together with fisheries information on elasmobranchs for assessment and conservation.

- Reducing the efforts of small-scale fisheries is required, especially in front of the Delta region. It causes mortality and threats to early-life stages of megafauna with respect to large rays and skates during pupping period (or calving period) .

- Mapping of hot spot areas for spawning and nursery is required for elasmobranchs that can be recommended as marine protected areas (MPAs).

- The handling procedures of bycatch are advisable to be operated faster, giving a chance for species that still live to be released into the water and in gentle style. It recommends using sharp and strong collapse to cut the hooks in the hard case from the outside mouth of sharks and rays to avoid damage to species.

- For purse seiners and trawlers, it is recommended to apply smooth metal channel inclined to the water to using it to release large sharks, rays and Sea turtles in easy-gently back to water with avoidance the damage and death onboard.

- Mitigation measures should be experimented for various gears and vulnerable species. For longline, the monofilament snoods is recommended, the sharks can be more easily cut it, and a circle hook is recommended by the ICCAT for Sea turtles. It is also recommended to follow the Guidelines of SPA/RAC (Action Plan for the Conservation of Cartilaginous Fish in the Mediterranean Sea), which recommend the use of circular hooks, where J-hooks are more likely to be swallowed than circular hooks.

- Stop returning the extracted marine litters during fishing activity back to water is recommended. This mitigates the impacts of nets remains and plastic pollution on biodiversity. It will be better to utilize and exploit the macroplastics via its recycling, this is agreed with the UNEP initiatives.

## Acknowledgements

The authors of this work present thanks to the authors of the published papers presented in this chapter. The appreciated thanks are presented to Ministry of Environment, Egypt, for the continuous initiatives and agreement to protect the environment and conserve the biodiversity. Also, many thanks go to the Agency (LFRPDA) which was recognized as GAFRD previously; and NIOF, Egypt; GFCM; ICCAT; SPA/RAC for attentions towards the conservation and protection of biodiversity particularly sharks and rays, sea turtles and cetaceans via recommendations and decrees such as ICCAT (2011); GFCM (2009, 2012); GAFRD (2009) (Decree159) for dolphin; GAFRD (2012) (Decree 444) for prohibiting the fishing and trading of sharks; and GAFRD (2012) (Decree151) for prohibiting fishing sea turtles as target and bycatch and recommending return it to water. Wormley thanks the rescue team in Alexandria with respect to her founder Mai hamada head of the team. The special thanks introduced to the environmental mangers for the positive efforts applied by Dr. Hussein Rashad Manager of the Ashtom El Gamil protectorate that located near port said and Damietta to rescue several sea turtles and return it to the sea in harmony with increase in the awareness among some fishers and traders hopping these situations will be considered in other areas particularly Alexandria and Abu Qir. Finally, special thanks go to Dr Lonbna Ben Nakhla, Programme Officer – Species Conservation, Tunisia (SPA/RAC) for financial support during the publishing this chapter as appositive cooperative to support such studies delt with conservation and management measures of the environment and biodiversity.

**Author details**

Mahmoud M.S. Farrag[1*], Ahmed A. Mustafa[2], Ismail M. Abdelazim[2] and Yassein A. Osman[3]

1 Marine Biology, Zoology Department, Al-Azhar University, Assiut, Egypt

2 Egyptian Environmental Affairs Agency (EEAA), Ministry of Environment, Egypt

3 National Institute of Oceanography and Fisheries (NIOF), Egypt

*Address all correspondence to: m_mahrousfarrag@yahoo.com and mahmoudfarrag42@azhar.edu.eg

## References

[1] Lejeusne C, Chevaldonné P, Pergent-Martini C, Boudouresque CF, Pérez T. Climate change effects on a miniature ocean: The highly diverse, highly impacted Mediterranean Sea. Trends in Ecology & Evolution. 2010;**25**:250-260

[2] Micheli F et al. Cumulative human impacts on Mediterranean and black sea marine ecosystems: Assessing current pressures and opportunities. PLoS One. 2013;**8**:e79889

[3] Farrag MS, Ahmed HO, TouTou MM, Eissawi MM. Marine mammals in the Egyptian Mediterranean coast "records and vulnerability". International Journal of Ecotoxicology and Ecobiology. 2019;**4**(1):8-16. DOI: 10.11648/j.ijee.20190401.12

[4] Tudela S. Ecosystem Effects of Fishing in the Mediterranean: An Analysis of the Major Threats of Fishing Gear and Practices to Biodiversity and Marine Habitats. Studies and Reviews, General Fisheries Commission for the Mediterranean. Vol. 74. Rome, Italy: Food and Agricultural Organization; 2004

[5] Sacchi J. Impact des techniques de peche sur l'environnement en Mediterranee. Etudes et revues. Commission Generale des Peches pour la Mediterranee. Vol. 84. Rome, Italy: Food and Agricultural Organization; 2008

[6] Ferreti F, Myers RA, Serena F, Loze HK. Loss of large predatory sharks from the Mediterranean Sea. Conservation Biology. 2008;**22**:952-964

[7] Casale P, Freggl D, Rocco M. Mortality induced by drifting longline hooks and branchlines in loggerhead sea turtles, estimated through observation in captivity. Aquatic Conservation: Marine and Freshwater Ecosystems. 2008;**18**:945-954

[8] Bearzi G. Interactions between cetaceans and fisheries in the Mediterranean Sea. Cetaceans of the Mediterranean and Black Seas: State of Knowledge and Conservation Strategies. di Notarbartolo Sciara G, 2002 9.1-9.20. ACCOBAMS, Monaco

[9] Gallucci VF, Taylor IG, Erzini K. Conservation and management of exploited shark populations based in reproductive value. Canadian Journal of Fisheries and Aquatic Sciences. 2006;**63**:931-942

[10] Dulvy NK, Fowler SL, Musick JA, Cavanagh RD, Kyne PK, Harrison LR, et al. Extinction risk and conservation of the world's sharks and rays. eLife. 2014;**3**:e00590

[11] Dulvy NK, Allen DJ, Ralph GM, Walls RHL. The Conservation Status of Sharks, Rays and Chimaeras in the Mediterranean Sea [Brochure]. Malaga, Spain: IUCN; 2016

[12] Hall MA, Alverson DL, Metuzals KI. By-catch: Problems and solutions. Marine Pollution Bulletin. 2000;**41**:204-219

[13] Worm B, Barbier EB, Beaumon N, Dufy JE, Folke C, Halpern BS, et al. Impacts of biodiversity loss on ocean ecosystem services. Science. 2006;**314**:787-790

[14] Stevens JD, Bonil R, Dulvy NK, Walker PA. The effects of fishing on sharks, rays, and chimaeras (chondrichthyes), and the implications for marine ecosystems. ICES Journal of Marine Science: Journal du Conseil. 2000;**57**:476-494

[15] Simpfendorfer CA, Heupel MR, White WT, Dulvy NK. The importance of research and public opinion to conservation management of sharks and rays: A synthesis. Marine and Freshwater Research. 2011;**62**:518-527

[16] Bradai MN, Saidi B, Enajjar S. Elasmobranchs of the Mediterranean and Black Sea: Status, Ecology and Biology. Bibliographic Analysis. Studies and Reviews. General Fisheries Commission for the Mediterranean. Vol. 91. Rome: FAO; 2012. p. 103

[17] Soykan C, Moore J, Zydelis R, Crowder L, Saina C, Lewison R. Why study by-catch? An introduction to the theme section on fisheries by-catch. Endangered Species Research. 2008;**5**:91-102

[18] Alfaro-Shigueto J, Mangel JC, Bernedo F, Duton PH, Seminof JA, Godley BJ. Small scale fisheries of Peru: A major sink for marine turtles in the Pacific. Journal of Applied Ecology. 2011;**48**:1432-1440

[19] Aldebert Y. Demersal resources of the Gulf of lions (NW Mediterranean). Impact of exploitation on fish diversity. Vie Milieu. 1997;**47**:275-284

[20] Jukic-Peladic S, Vrgoc N, Krstulovic-Sifner S, Piccineti C, Piccineti-Manfrin G, Marano G, et al. Long–term changes in demersal resources of the Adriatic Sea: Comparison between trawl surveys carried out in 1948 and 1998. Fisheries Research. 2001;**53**:95-104

[21] Ferreti F, Myers RA, Serena F, Sartor P. Long term dynamics of chondrichthyan ish community in the upper Tyrrhenian Sea. In: ICES Council Meeting Documents. Copenhagen: ICES CM; 2005. p. 25

[22] Bradai MN, Saidi B, Enajjar S. Overview on Mediterranean Shark's Fisheries: Impact on the Biodiversity (in Marine Ecology - Biotic and Abiotic Interactions). London, UK: Intech Open; 2018. DOI: 10.5772/intechopen.74923

[23] Ragheb E, Hasan MWA. First record of *Pteroplatytrygon violacea* (Bonaparte, 1832) with annotation of cartilaginous fishes by-catch by gill nets (Egyptian Mediterranean). Egyptian Journal of Aquatic Research. 2021;**47**(4):387-392

[24] Ibrahim MA, Hassan MAW, El-Far AMM, Farrag EFE, Farrag MMS. Deep sea shrimp resources in the south eastern Mediterranean waters of Egypt. Egyptian Journal of Aquatic Research. 2011;**37**(2):131-137

[25] Farrag MMS. Deep-sea ichthyofauna from eastern Mediterranean Sea, Egypt: Update and new records. Egyptian Journal of Aquatic Research. 2016;**42**(4):479-489

[26] Farrag MMS. An evaluation of the deep-sea catch in the Mediterranean Sea, Egypt regarding pattern of CPUE, diversity, sharks, and discards. Scientific African. 2022;**18**:e01431. DOI: 10.1016/j.sciaf.2022.e01431

[27] Farrag MMS, Adel M. Diversity, Distribution, and Exploitation of Cartilaginous Fish along the Egyptian Mediterranean Coast: Conservation and Bycatch Challenges; 2025. (in press)

[28] Frazier J. Prehistoric and ancient historic interactions between humans and marine turtles. In: Lutz PL, Musick JA, Wyneken J, editors. The Biology of Sea Turtles. Volume II, CRC Marine Biology Series. Boca Raton, London, New York, Washington D.C.: CRC Press, Inc.; 2003. pp. 1-38

[29] National Research Council. Decline of Sea Turtles: Causes and Prevention. Washington, DC, USA: National Research Council; 1990

[30] Casale P, Freggl D, Gratton P, Argano R, Oliverio M. Mitochondrial DNA reveals regional and interregional importance of the central Mediterranean African shelf for loggerhead sea turtles (Caretta caretta). Scientia Marina. 2008;**72**(3):541-548. DOI: 10.3989/scimar.2008.72n3541

[31] Fleming EH. Swimming against the Tide: Recent Surveys of Exploitation, Trade, and Management of Marine Turtles in the Northern Caribbean. Washington, DC, USA: TRAFFIC North America; 2001

[32] Campbell LM. Contemporary culture, use, and conservation of sea turtles. In: Lutz PL, Musick JA, Wyneken J, editors. The Biology of Sea Turtles. Volume II, CRC Marine Biology Series. Boca Raton, London, New York, Washington D.C.: CRC Press, Inc.; 2003. pp. 307-338

[33] IUCN. Marine Mammals and Sea Turtles of the Mediterranean and Black Seas Gland. Switzerland and Malaga, Spain: IUCN; 2012. 32 p

[34] Lewison RL, Crowder LB, Read AJ, Freeman SA. Understanding impacts of fisheries bycatch on marine megafauna. Trends in Ecology and Evolution. 2004;**19**:598-604

[35] Campell LM, Silver JJ, Gray NJ, Ranger S, Broderick AC, Fisher T, et al. Co-management of sea turtle fisheries: Biogeography versus geopolitics. Marine Policy. 2009;**33**:137-145

[36] Mancini A, Koch V. Sea turtle consumption and black-market trade in Baja California Sur, Mexico. Endangered Species Research. 2009;**7**:1-10

[37] Broderick AC, Glen F, Godley BJ, Hays GC. Estimating the size of nesting populations of green and loggerhead turtles in the Mediterranean. Oryx. 2002;**36**:227-236

[38] Laurent L, Abd El-Mawla EM, Bradai MN, Demirayak F, Oruc A. Reducing Sea Turtle Mortality Induced by Mediterranean Fisheries: Trawling Activity in Egypt, Tunisia and Turkey. Rome, Italy: Report for the WWF International Mediterranean Programme. WWF Project 9E0103; 1996

[39] Nada M, Casale P. Sea turtle bycatch and consumption in Egypt threatens Mediterranean turtle populations. Fauna & Flora International, Oryx. 2011;**45**(1):143-149. DOI: 10.1017/S0030605310001286

[40] Clarke M, Campbell AC, Hameid WS, Ghoneim S. Preliminary report on the status of marine turtle nesting populations on the Mediterranean coast of Egypt. Biological Conservation. 2000;**94**:363-371

[41] Nada MA. Observations on the trade in sea turtles at the fish market of Alexandria, Egypt. Zoology in the Middle East. 2001;**24**(109):118

[42] Venizelos L, Nada M. Exploitation of loggerhead and green turtles in Egypt: Good news? Marine Turtle Newsletter. 2000;**87**:12-13

[43] Broderick AC, Coyne MS, Fuller WJ, Glen F, Godley BJ. Fidelity and overwintering of sea turtles. Proceedings of the Royal Society B. 2007;**274**:1533-1538

[44] Rees AF, Jony M, Margaritoulis D, Godley BJ. Satellite tracking of a green turtle Chelonia mydas from Syria further highlights importance of North Africa for Mediterranean turtles. Zoology in the Middle East. 2008;**45**:49-54

[45] Margaritoulis D, Argano R, Baran I, Bentivegna F, Bradai MN, Caminas JA,

et al. Loggerhead turtles in the Mediterranean Sea: Present knowledge and conservation perspectives. In: Bolten AB, Witherington B, editors. Loggerhead Sea Turtles. Washington, DC, USA: Smithsonian Institution Press; 2003. pp. 175-198

[46] Farrag MMS, TouTou MMM, Eissawi MM, Besar MH, Salama AF, Ahmed HO. Re-identification of Bryde's whale (*Balaenoptera edeni*) and Gervais' beaked whale (*Mesoplodon europaeus*) on the Mediterranean coast of Egypt updating, strandings in opposite to climatic and anthropogenic impacts. Egyptian Journal of Aquatic Biology And Fisheries. 2022;**26**(6):125-153

[47] Frantzis A, Airoldi S, Notarbartolo-di-Sciara G, Johnson C, Mazzarri S. Inter-basin movements of Mediterranean sperm whales provide insight into their population structure and conservation. Deep Sea Research. 2011;**I**(58):454-459

[48] Panigada S et al. Estimating cetacean density and abundance in the central and Western Mediterranean Sea through aerial surveys: Implications for management. Deep Sea Research. 2017;**Part II 141**:41-58

[49] Frantzis A, Alexiadou P, Paximadis G, Politi E, Gannier A. Current knowledge of the cetacean fauna of the Greek seas. Journal of Cetacean Research and Management. 2003;**5**:219-232

[50] David L, Alleaume S, Guinet C. High risk areas of collision between fin whales and ferries in the North-Western Mediterranean sea. Journal of Marine Animal and Their Environment. 2011;**4**:17-28

[51] Castellote M, Clark CW, Lammers MO. Acoustic and behavioural changes by fin whales (*Balaenoptera physalus*) in response to shipping and airgun noise. Biological Conservation. 2012;**147**:115-122

[52] Caadas A, Vzquez JA. Conserving Cuvier's beaked whales in the Alboran Sea (SW Mediterranean): Identification of high-density areas to be avoided by intense man-made sound. Biological Conservation. 2014;**178**:155-162

[53] Bearzi G, Agazzi S, Gonzalvo J, Costa M, Bonizzoni S, Politi E, et al. Overfishing and the disappearance of short-beaked common dolphins from western Greece. Endangered Species Research. 2008;**5**:1-12. DOI: 10.3354/esr00103

[54] Öztürk AA, Tonay AM, Dede A. Review article strandings of the beaked whales, Risso's dolphins, and a minke whale on the Turkish coast of the eastern Mediterranean Sea. Journal of Black Sea/Mediterranean Environment. 2011;**17**(3):269-274

[55] Maio N, Pollaro F, Di Nocera F, De Carlo E, Galiero G. *Cetacei spiaggiati* lungo le coste della Campania dal 2006 al 2011 (Mammalia: Cetacea). Atti della Società italiana di scienze naturali e del Museo civico di storia naturale di Milano. 2012;**153**:241-255

[56] Goffman O, Roditi M, Shariv T, Spanier E, Kerem D. Cetaceans from the Israeli coast of the Mediterranean Sea. Israel Journal of Zoology. 2000;**46**:143-147

[57] Farrag MMS. Fisheries and Biological Studies on Lessepsian Pufferfish, Lagocephalus sceleratus (Gmelin, 1789) (Family: Tetraodontidae) in the Egyptian Mediterranean Waters [Thesis]. Egypt: Faculty of Science Al-Azhar University Boys Assiut; 2014

[58] Abo-Taleb HA, El-feky MMM, El-Tabakh MAM, Hendy DM,

Maaty M. First record of Bryde's whale (*Balaenoptera brydei*, Olsen, 1913) in the southeastern Mediterranean Sea, Alexandria, Egypt. Egyptian Journal of Aquatic Biology and Fisheries. 2020;**24**(1):667-695

[59] Shirihai H, Jarrett B. Whales, Dolphins, and Other Marine Mammals of the World. Princeton Field Guides. Princeton, NJ: Princeton University Press; 2006. pp. 43-45. ISBN 978-0-691-12757-6

[60] Jefferson TA, Webber MA, Pitman RL. Marine Mammals of the World: A Comprehensive Guide to their Identification. London: Academic Press; 2008. p. 573

[61] Cerchio S, Andrianantenaina B, Lindsay A, Rekdahl M, Andrianarivelo N, Rasoloarijao T. Omura's whales (*Balaenoptera omurai*) off Northwest Madagascar: Ecology, behaviour and conservation needs. Royal Society Open Science. 2015;**2**:150301

[62] Ranjbar S, Dakhteh MS, Waerebeek KV. Omura's whale (*Balaenoptera omurai*) stranding on Qeshm Island, Iran: Further evidence for a wide (sub)tropical distribution, including the Persian Gulf. Journal of Marine Biology and Oceanography. 2016;**5**:1-9

[63] de Vos A. First record of Omura's whale, *Balaenoptera omurai*, in Sri Lankan waters. de Vos Mar Biodiversity Records. 2017;**10**:18. DOI: 10.1186/s41200-017-0121-2

[64] Serena F. Field Identification Guide to the Sharks and Rays of the Mediterranean and Black Sea. FAO Species Identification Guide for Fishery Purposes. Rome: FAO; 2005. p. 97

[65] FAO. Information on Fisheries Management in the Arab Republic of Egypt. 2003. Available from: http://www.fao.org/fi/fcp/en/EGY/body.htm [Accessed: September 17, 2007]

[66] UNEP/MAP - SPA/RAC. Action Plan for the Conservation of Marine Turtles in the Mediterranean. Ed. SPA/RAC, Tunis: 16 pp + Annexes. 2019

[67] GFCM. GFCM/42/2018/2 on Fisheries Management Measures for the Conservation of Sharks and Rays in the GFCM Area of Application, Amending Recommendation GFCM/36/2012/3. GFCM. Rome: FAO; 2018

[68] SPA/RAC–UN Environment/MAP. In: Bradai MN, editor. Action Plan for the Conservation of Cartilaginous Fishes (Chondrichtyans) in the Mediterranean Sea. Tunis: SPA/RAC; 2020. 18 p

[69] Ebert DA, Stehmann MFW. Sharks, batoids, and chimaeras of the North Atlantic. In: FAO Species Catalogue for Fishery Purposes. Vol. 7. Rome: FAO; 2013. 523 p

[70] Otero M, Serena F, Gerovasileiou V, Barone M, Bo M, Arcos JM, et al. Identification Guide of Vulnerable Species Incidentally Caught in Mediterranean Fisheries. Malaga: IUCN; 2019

[71] Serena F, Abella AJ, Bargnesi F, Barone M, Colloca F, Ferretti F, et al. Species diversity, taxonomy, and distribution of Chondrichthyes in the Mediterranean and Black Sea. The European Zoological Journal. 2020;**Vol. 87**(1):497-536. DOI: 10.1080/24750263.2020.1805518

[72] Chuang P-S, Hung T-C, Chang H-A, Huang C-K, Shiao J-C. The species and origin of shark fins in Taiwan's fishing ports, markets, and customs detention: A DNA barcoding analysis. PLoS One. 2016;**11**(1):e0147290. DOI: 10.1371/journal. pone.0147290

[73] Podestà M, D'Amico A, Pavan G, Drougas A, Komnenou A, Portunato N. A review of Cuvier's beaked whale strandings in the Mediterranean Sea. Journal of Cetacean Research and Management. 2006;7:251-261

[74] Wells RS, Scott MD. Common bottlenose dolphin *Tursiops truncatus*. In: Perrin WF, Würsig B, Thewissen JGM, editors. Encyclopedia of Marine Mammals. 2nd ed. San Diego, CA, USA: Academic Press; 2009. pp. 249-255

[75] Notarbartolo di Sciara G, Kerem D, Smeenk C, Rudolph P, Cesario A, Costa M, et al. Cetaceans of the Red Sea. CMS Technical Series. 2017;**33**:1-86

[76] Reeves R, Notarbartolo di Sciara G. The Status and Distribution of Cetaceans in the Black Sea and Mediterranean Sea IUCN Centre for Mediterranean Cooperation. Malaga, Spain; 2006. 137 p

[77] Giovanni B, Caterina MF, Randall RR. Ecology and conservation of common bottlenose dolphins *Tursiops truncatus* in the Mediterranean Sea. Mammal Review. 2008;**39**(2):92-123

Section 2

# Threats Impacting Endangered Marine Vertebrates and Environment

Chapter 4

# Effects of Marine Threats on Mediterranean Marine Top Predators

*Danilo Malara and Cristina Pedà*


## Abstract

This chapter explores the growing threats faced by top predators in the Mediterranean Sea. Climate change, overfishing, and plastic pollution may affect their migration patterns and lead them to abandon established reproduction areas. For instance, tuna might not find its spawning grounds due to warmer water, sea turtles might fight against plastic debris to lay their eggs, or dolphins might be tangled by a ghost net. These disruptions threaten entire predator populations, posing at risk the entire species with the death of young specimens and the reduction of offspring. The entire marine ecosystem suffers as these top predators play a vital role in maintaining a healthy balance. By examining the impact of different threats on biological and ecological aspects of movement and reproduction, this chapter shows the vulnerabilities of these important creatures, highlighting the urgent need for conservation efforts. Only with immediate actions can we ensure their survival and a healthy Mediterranean Sea for generations to come.


**Keywords:** Mediterranean Sea, iconic marine vertebrates, marine litter, ecosystem conservation, climate change, fishery

## 1. Introduction

The Mediterranean Sea, the origin of numerous empires and cultures and a vibrant hub of marine life, is facing a hidden crisis. This precious body of water, enclosed by diverse coastlines and steeped in history, is more fragile than it seems. Unlike the vast open oceans, the Mediterranean Sea is a semi-enclosed basin, connected to the Atlantic Ocean only through the narrow Strait of Gibraltar [1], to the Red Sea by the Suez channel [2], and to the Black Sea through the Dardanelles Straits. This limited exchange of water, with a residence time of approximately 80–150 years [3], makes it particularly vulnerable to environmental stressors. This means pollution increases, temperatures rise faster, and any environmental change has a large impact.

The Mediterranean's water circulation is a dynamic and intricate system shaped by a complex interplay of wind patterns, temperature gradients, salinity variations, the Earth's rotation, and the basin's topography. In general, less dense and warmer surface waters flow eastward from the Atlantic through the Strait of Gibraltar, carrying nutrients and oxygen into the basin [3]. This inflow is counterbalanced by a westward flow of deeper, saltier waters, which exit through the strait's depths [4]. This thermohaline circulation, driven by differences in temperature and salinity, is crucial for maintaining the Mediterranean's overall health and ecological balance. Additionally, wind-driven upwelling significantly influences the distribution of marine organisms in the Mediterranean Sea, creating productive fishing grounds in specific nutrient-rich areas [5]. However, the delicate balance of this circulation is sensitive to disruptions, particularly those induced by climate change and human activities. For example, the increasing temperatures in the Mediterranean are leading to a stronger stratification of the water column, with a distinct separation between warmer surface waters and cooler deep waters [6]. This stratification can hinder the mixing of nutrients and oxygen, potentially impacting the entire food web. Additionally, the semi-enclosed nature of the Mediterranean Sea means that any changes in the Atlantic inflow, such as variations in temperature or salinity, can have a huge effect on the sea's circulation and overall health.

Understanding these complex dynamics is crucial for predicting the future of the Mediterranean Sea and developing strategies to mitigate the impacts of climate change and other environmental stressors. Sadly, this intricate system is increasingly threatened by climate change, overfishing, and the persistent drift of plastic pollution, which are altering the sensitive balance of the Mediterranean ecosystem. These threats are not just numbers on a chart or abstract concepts, but the devastating consequences for these key species are tangible.

## 2. Climate change: Heatwaves and rise temperature

Ocean temperatures are crucial for various biological functions, significantly impacting an animal's metabolic rate [7, 8]. This impact, in turn, plays a role in shaping various aspects of the animal's life, including its distribution and behavior patterns. For instance, different animals such as billfish [9], sharks [10], and mammals [11] dive deep into the mesopelagic zone, sometimes venturing below 200 meters. This behavior can serve a variety of purposes, such as escaping predators or finding food. However, recent research on the Mediterranean spearfish (*Tetrapturus belone*) has shown that seasonal changes in the thermocline in the Mediterranean Sea—the boundary between warmer surface water and cooler, deeper water—influence their diving behavior, leading them to dive deeper during the fall and winter [12]. This shift in diving behavior coincides with the thermocline's seasonal deepening in the Mediterranean Sea [13]. Those seasonal variations significantly influence the vertical distribution of prey, essentially affecting the feeding habits and behavior of large pelagic animals [14].

These shifts in the depth of the mixed layer within the Mediterranean Sea play a crucial role in determining the overall temperature structure of the water column. Billfish, in particular, seem to adjust their vertical movements in response to changing water temperatures [9]. They do this to maximize their chances of finding food in the parts of the water column where their bodies can function best [15]. However, these fluctuations in the mixed layer depth can have a major impact on the behavior and preferred depths of top predators such as *Thunnus thynnus* (Atlantic bluefin tuna) [16]. Indeed, the Atlantic bluefin tuna are vulnerable to temperature changes and likely will pay the price. The Mediterranean is warming at an alarming rate, 20%

faster than the global average, according to the Intergovernmental Panel on Climate Change [17]. This warming trend disrupts the sensitive balance of the ecosystem, leading to cascading effects that threaten its top predators. The effect can be summarized in three main issues:

### 2.1 Shifting habitats

The warming of the Mediterranean Sea is driving a phenomenon known as "tropicalization." This means that species are being forced to migrate in search of cooler waters that better suit their needs, altering the intricate balance of the ecosystem [18]. This shift is not only changing where species are found but also how many there are and how they interact with each other. For example, recent research has shown that the Atlantic bluefin tuna (*T. thynnus*) in the Mediterranean could move further north than usual due to rising sea temperatures [19]. This northward shift not only disrupts traditional fishing patterns but also raises concerns about the long-term sustainability of the tuna population. On the other hand, the swordfish (*Xiphias gladius*) is also experiencing changes in its migration patterns and distribution due to shifting ocean currents and warming waters [20]. These changes are impacting the swordfish's ability to reproduce successfully, as their breeding grounds and access to food sources are altered. In some species, temperature is also important for sex differentiation during development. Therefore, warmer sand temperatures due to climate change are leading to nesting disruptions and female-biased sex ratios in sea turtles, including both loggerhead (*Caretta caretta*) and green sea turtles (*Chelonia mydas*), nesting in the Mediterranean Sea and other regions [21, 22]. On the other hand, the impacts of rising sea levels on the endangered Mediterranean fin whale (*Balaenoptera physalus*) are complex [23]. While the loss of coastal habitats could negatively impact the population, the opening of new areas due to melting sea ice may offer some benefits [24], highlighting the intricate and unpredictable consequences of climate change on marine ecosystems. However, these potential gains are coupled with uncertainties about prey availability and the whales' adaptability to a changing environment. The overall impact of climate change on this vulnerable population remains intricate and unpredictable (**Figure 1**).

### 2.2 Acidification

The increasing absorption of atmospheric carbon dioxide by the Mediterranean Sea is driving a concerning rise in its acidity [25]. This phenomenon, termed ocean acidification, poses a significant threat to the marine ecosystem, particularly its top predators. Shell-forming organisms like oysters and mussels, vital components of the marine food web, are especially vulnerable to acidification. Their shells become thinner and more fragile as carbonate ions, essential for shell formation, become less available in the increasingly acidic environment [25]. This compromises their survival and reduces their abundance, impacting the predators that rely on them for sustenance [26]. Additionally, acidification can directly impact the development and survival of fish larvae, including those of large pelagic predators like sharks [27]. The development and survival of fish larvae are hindered in acidic conditions, potentially leading to population declines and disruptions in the predator-prey dynamics of the ecosystem. The consequences for the Mediterranean's apex predators are huge. Reduced prey availability, coupled with the challenges faced by their offspring, can lead to nutritional stress, decreased reproductive success, and ultimately, population declines (**Figure 2**) [28].

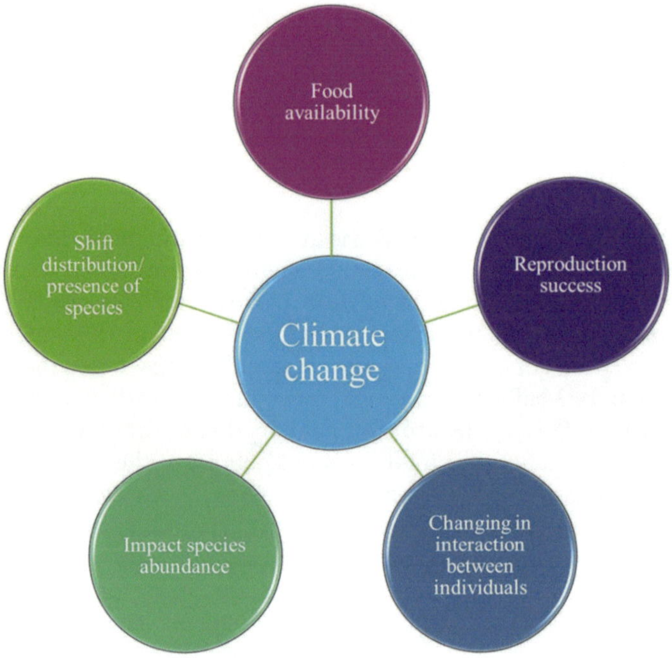


**Figure 1.**
*Effects of climate change on marine top predators.*

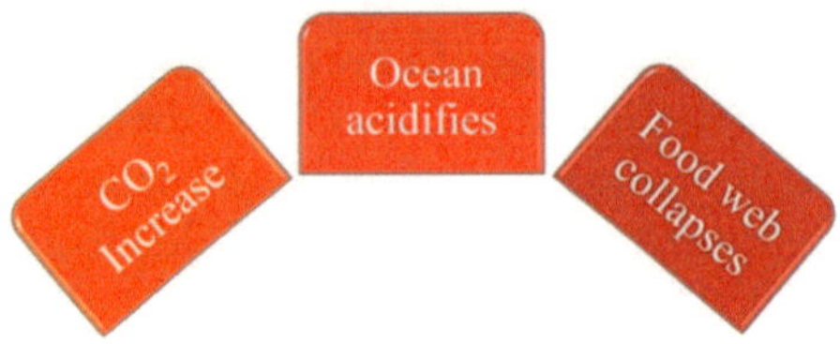


**Figure 2.**
*Effect of increasing the $CO_2$.*

The marine ecosystem is under siege, with the potential for long-term ecological imbalances and cascading effects throughout the food web. Addressing ocean acidification requires urgent action to mitigate climate change and reduce carbon emissions. Protecting the Mediterranean's top predators and preserving the health of this vital marine ecosystem necessitate a concerted global effort.

## 2.3 Extreme weather

Rising global temperatures are having a severe impact on marine ecosystems worldwide, but the enclosed nature and shallow depths of the Mediterranean Sea make it exceptionally susceptible to the effects of climate change [29]. Among the most pressing concerns are the intensifying heatwaves, both terrestrial and marine, and the gradual but persistent rise in sea temperatures. These tangled phenomena set off a cascade of ecological disruptions, jeopardizing the survival of the Mediterranean's apex predators and the stability of its marine ecosystem as a whole. The Mediterranean is experiencing an unambiguous increase in the frequency and

*DOI: http://dx.doi.org/10.5772/intechopen.1007420*

intensity of heatwaves [30]. Complex atmospheric patterns, amplified by anthropogenic greenhouse gas emissions, are responsible for these extended periods of extreme heat [31]. In the marine environment, heatwaves disrupt vital ocean currents, cause mass mortality events among sensitive species like corals, and create conditions that are increasingly inhospitable for a wide range of marine life. The steady increase in sea surface temperatures, a direct result of global warming, amplifies the effects of heatwaves [32]. The Mediterranean Sea, being enclosed and relatively shallow, absorbs and retains significant amounts of heat, leading to a worrying trend of rising average temperatures [33]. This warming trend disrupts the delicate equilibrium of marine ecosystems, with ramifications extending from the base of the food web to its apex predators. The warming waters of the Mediterranean are causing shifts in the distribution and abundance of prey species. Small fish, crustaceans, and other organisms that constitute the foundation of the food web for apex predators may become scarce or migrate to different areas in search of cooler waters. This leaves top predators like sharks, tuna, and dolphins struggling to find sufficient sustenance, with potentially devastating consequences [34]. When prey becomes scarce, top predators struggle to meet their energetic demands. This can result in malnutrition, hindering their overall health and vitality. In severe cases, prolonged food shortages can lead to starvation and death, impacting population numbers and potentially disrupting the entire food web [1]. Insufficient food intake can negatively impact reproductive capabilities. Starving individuals may have reduced fertility, produce fewer offspring, or be unable to provide adequate care for their young, leading to lower recruitment rates and population decline [35]. Prey scarcity can trigger changes in predator behavior. Predators might intensify competition with each other, shift to less preferred prey species, or even resort to cannibalism in extreme cases. These altered dynamics can have cascading effects on the entire ecosystem. For instance, the loss of predation risk

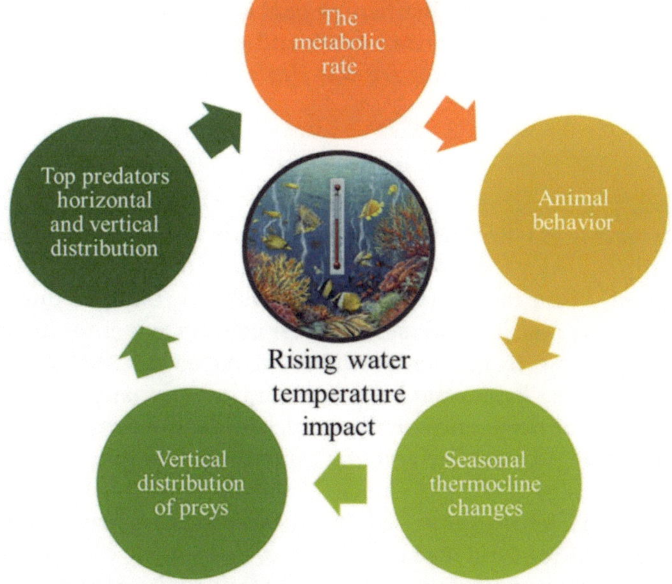


**Figure 3.**
*Schematic representation of the effects of global rising temperature.*

from apex predators can amplify the effects of marine tropicalization, particularly following extreme climatic events [36]. Such changes could lead to further imbalances in Mediterranean marine ecosystems, exacerbating the consequences of prey depletion [37]. These altered dynamics can have cascading effects on the entire ecosystem. These altered dynamics can have cascading effects on the entire ecosystem. The cascading effects of these changes echo throughout the ecosystem, potentially altering predator-prey dynamics, trophic interactions, and overall ecosystem stability (**Figure 3**).

Addressing this crisis requires a multifaceted approach. Ongoing research and monitoring are crucial to understand the complex interplay between climate change, heatwaves, and the marine ecosystem [38]. Conservation strategies, such as the establishment of marine protected areas and the implementation of sustainable fishing practices, can help buffer the impacts on vulnerable species. However, the most critical step is to curb greenhouse gas emissions at a global scale [17]. The fate of the Mediterranean, its top predators, and the intricate web of life it supports hinges on our collective ability to mitigate climate change before irreversible damage is done.

## 3. Marine litter

In recent decades, marine litter (ML) pollution has emerged as a pressing global concern due to its far-reaching implications for the marine environment and its inhabitants. This escalating issue has prompted extensive research and advocacy efforts [39, 40], highlighting the urgent need for action. The primary driver of ML's omnipresence across diverse marine habitats, from coastlines and the open ocean to the seafloor, is the increasing intensity of human activities both on land and at sea [41]. Population growth, coastal development, industrialization, and expanded maritime activities like shipping and tourism have all contributed to the surge of debris entering our oceans. ML encompasses a wide range of materials, with plastics being the most prevalent, including derelict fishing gear and single-use plastics (SUP). However, it is also about rubbers, glass, metal, and various other discarded items. The impacts of ML on marine ecosystems are complex and severe.

### 3.1 Ingestion

ML, especially plastic debris, poses a serious threat to marine life through ingestion. Plastics in marine environments may derive from the direct introduction of microbeads or from the degradation of large plastics by the action of abiotic and (sun, seawater, and wind) biotic factors (microorganisms). Due to its chemical and physical properties, plastic may distribute along the water column and enter the trophic web. A vast array of organisms, from microscopic zooplankton to large marine vertebrates, according to their feeding habits and behavior, may intentionally ingest litter, often mistake plastic items for prey, accidentally eat debris during foraging activity, as well as ingest contaminated prey [42]. Ingested plastics may cause physical harm and, also, promote the transfer of toxic chemicals (POPs and plasticizer) in organisms and in the trophic web [43, 44]. Ingested plastic debris can obstruct digestive tracts, hinder nutrient absorption, and lead to blockages, malnutrition, impaired growth and reproduction, and even mortality [45, 46]. In the Mediterranean Sea, all three species of sea turtles, loggerhead (*C. caretta*), green (*C. mydas*), and particularly the leatherback (*Dermochelys coriacea*), frequently mistake plastic bags for jellyfish, one of their main prey [47]. This leads to plastic accumulation in their gut, causing blockages, lethargy, and ultimately death [48]. Plastic remains have also been found in the stomachs of commercially valuable fish species, raising concerns about the transfer of contaminants up the food chain and to humans [49, 50]. This can have important consequences, particularly for apex predators in vulnerable ecosystems like the Mediterranean basin.

### 3.2 Ghost nets and gears

Lost or discarded fishing gear, often referred to as "ghost gear," poses a substantial and ongoing threat to marine ecosystems worldwide. This debris, including fishing lines, nets, ropes, and other synthetic materials, persists in the environment for extended periods, interacting with marine life in detrimental ways. Entanglement in ghost gear can cause significant physical trauma, including lacerations, abrasions, and constrictions that can lead to tissue death or loss of limbs [39]. Entangled animals often experience restricted mobility, hindering their ability to hunt, evade predators, and migrate, which can lead to starvation and malnutrition [40]. For air-breathing marine animals, entanglement can be particularly deadly, preventing them from surfacing to breathe and resulting in suffocation and drowning [39]. Addressing the persistent issue of ML and its associated entanglement risks necessitates a multifaceted approach that targets various stages of the problem, from prevention to remediation.

The pervasive issue of ML demands urgent and comprehensive action. By implementing a combination of preventive measures, innovative technologies, and collaborative efforts, we can mitigate the impacts of ML on marine ecosystems and safeguard the health and well-being of marine life for future generations.

## 4. Human induced pressure

Fishing has been deeply connected with the cultures and economies of Mediterranean coastal communities for centuries [51]. The sea's rich biodiversity has traditionally supported livelihoods and provided sustenance. However, advancements in fishing technologies and the ever-increasing global demand for seafood have drastically intensified fishing pressure in recent times [52]. A wide array of fishing methods, including trawling, purse seining, and longlining, are employed to target diverse species, ranging from small pelagic fish to apex predators [53]. Unfortunately, the current scale of fishing in the Mediterranean has far surpassed sustainable levels, resulting in a phenomenon known as overfishing. This occurs when fish populations are exploited at a rate exceeding their natural capacity to replenish [54]. The consequences of overfishing move throughout the marine ecosystem, leading to the depletion of targeted fish stocks and the disruption of the entire food web [55]. As key prey species become scarce, top predators like sharks and dolphins face declining food sources and struggle to maintain their populations [56]. Additionally, overfishing can cause unintended harm to non-target species through bycatch, where animals such as sea turtles and dolphins are inadvertently captured in fishing gear.

Predation in the wild is a natural process governed by ecological balance. Unlike human activities like fishing or hunting, wild predators typically do not decimate entire prey populations. Instead, their predation rate is generally regulated by the availability of prey and their own metabolic needs, resulting in a dynamic equilibrium where both predator and prey populations can persist [57]. Predators (especially opportunistic ones such as sharks) primarily target individuals that are easier to catch, such as the young, old, sick, or injured. This selective predation helps to maintain the overall health and genetic diversity of the prey population [58]. Additionally, the fear of predation influences the behavior and distribution of prey species, shaping the overall structure of the ecosystem [59]. While there are instances where predator populations can exert significant pressure on prey populations, leading to temporary fluctuations, these events are typically part of natural cycles and rarely result in the complete eradication of a prey species [60].

When a top predator is exploited, the equilibrium of the environment is altered, where mesopredators or other opportunistic species increase in number, causing issues in the balance of the environment. Significant declines in apex predators such as sharks, tuna, and billfish have been widely documented in the Mediterranean [56, 61], and their decline in the Mediterranean Sea has been linked to a notable increase in the abundance of certain mesopredators. For example, studies analyzing long-term trends in the Adriatic Sea have reported a significant reduction in the biomass of large-bodied sharks, rays, and skates, coinciding with an increase in the abundance of smaller elasmobranchs and teleosts [62]. This observation supports the notion of a trophic cascade triggered by the removal of apex predators, potentially leading to mesopredator release and subsequent ecosystemwide repercussions. Similarly, ecosystem modeling studies conducted in the Mediterranean Sea have predicted significant shifts in community structure following the decline of top predators [63]. The loss of these apex consumers can lead to cascading effects throughout the food web, impacting not only mesopredator populations but also lower trophic levels and overall ecosystem functioning. While the specific impacts of individual apex predator declines may vary depending on their ecological roles and dietary overlaps, the collective loss of multiple apex predators is likely to have more pronounced and far-reaching consequences for the Mediterranean marine environment. These findings emphasize the critical need for effective conservation measures to protect apex predators and maintain the delicate balance of this vital ecosystem.

Fisheries management must ensure that maximum sustainable yield (MSY) targets are not set too high, as this can lead to overfishing. However, it is not always easy to gather all the necessary information to create models using historical data [64]. Therefore, it's often necessary to make difficult decisions based on limited data and imperfect models. The work of a fisheries manager is crucial and requires them to act quickly and decisively, even when faced with opposition from stakeholders [64].

## 5. Conclusions

The Mediterranean Sea is facing an unprecedented ecological crisis. The synergistic effects of climate change, marine litter, and overfishing have triggered a cascade of disruptions that threaten the very foundation of this unique marine ecosystem. Rising temperatures and ocean acidification are reshaping the Mediterranean's ecological landscape, forcing species to migrate, disrupting reproductive cycles, and altering the availability of prey. Marine litter, particularly plastic debris and ghost gear, poses a direct threat to marine life through ingestion and entanglement, further stressing populations already struggling to adapt to a changing environment. Overfishing, intensified by technological advancements and unsustainable practices, is decimating fish stocks and disrupting the intricate food web, with cascading effects on apex predators and the ecosystem as a whole. The decline of these top predators can trigger trophic cascades, leading to mesopredator release and potential imbalances throughout the marine environment. Addressing this crisis demands a concerted global effort.

Addressing this crisis demands a concerted global effort. Mitigating climate change through reduced carbon emissions is paramount, as it underpins many of the challenges faced by the Mediterranean. Implementing robust waste management practices to reduce marine litter at its source and transitioning to sustainable fishing practices that prioritize ecosystem health are also crucial steps. Investing in research and monitoring will be key to understanding the complex interplay of these stressors and informing effective conservation strategies.

The challenges are immense, but the stakes are even higher. The future of the Mediterranean Sea, its iconic apex predators, and the countless communities that depend on its health hang in the balance. The Mediterranean Sea is a hotspot of biodiversity, harboring a remarkable array of marine life, many of which are found nowhere else on Earth. Preserving this natural heritage is not only an ecological imperative but also a cultural and economic one. Only through decisive action and international cooperation can we hope to safeguard this unique ecosystem and ensure its resilience for generations to come.

A comprehensive strategy to protect and address the complex challenges facing the Mediterranean Sea and ensure its long-term health and resilience must include:

- *Monitoring studies based on climate change*: Continuous monitoring and research on the impacts of climate change on the Mediterranean ecosystem are essential for adapting conservation strategies and predicting future challenges. This includes studying changes in species distribution, ocean temperature, and acidity levels.

- *Stock conservation*: Protecting and restoring fish stocks is crucial for ensuring the long-term viability of the Mediterranean's marine life. This includes measures such as establishing marine protected areas, envisaging fishing quotas, and combating illegal, unreported, and unregulated (IUU) fishing. While it is not related to the actual threat, it prevents the complete exploitation of the fishing resources.

- *Sustainable consumption*: Encouraging responsible seafood consumption habits can help reduce pressure on overexploited fish stocks and promote sustainable fishing practices. Consumer choices can play a significant role in driving market demand for sustainably sourced seafood.

- *Improved waste management practices*: Enhancing waste management practices on land is fundamental to reducing marine litter. This includes promoting proper waste disposal and recycling programs, raising public awareness, and encouraging individual responsibility.

- *Reduction of fishing gear loss*: Implementing measures to minimize fishing gear loss is crucial. This can be achieved by promoting responsible fishing practices, educating fishers, encouraging gear retrieval and reporting, and incentivizing gear recovery through deposit-refund systems or buy-back programs.

- *Development of innovative gear technologies*: Investing in research and development of fishing gear that is less prone to loss or entanglement, such as biodegradable or easily detectable materials, can further contribute to reducing the problem.

- *Retrieval of ghost gear*: Organized efforts are needed to remove lost and abandoned fishing gear from the marine environment. Regular clean-up initiatives, utilizing advanced technologies like remotely operated vehicles (ROVs) and side-scan sonar, and collaborating with fishing communities can all contribute to effective ghost gear retrieval.

- *Development of alternative fishing gear*: Exploring and promoting the use of fishing gear that is less likely to cause entanglement, such as biodegradable fishing gear or gear with modifications to reduce entanglement potential, is a proactive approach to reducing risks.

- *Sustainable fisheries*: Implementing sustainable fishing practices, such as adhering to catch limits, using selective fishing gear, and protecting critical habitats, is vital for maintaining healthy fish populations and preserving the delicate balance of the marine ecosystem. The GFCM 2030 Strategy emphasizes the importance of sustainable fisheries and aquaculture as pillars for the livelihoods of coastal communities in the Mediterranean and Black Sea [65].

Although the ecological restoration of the Mediterranean Sea presents a formidable challenge, its future trajectory is not fixed. Through collaborative and innovative interventions, we can navigate toward a state of enhanced ecosystem health and resilience, thereby safeguarding its abundant resources and esthetic value for posterity.

## Acknowledgements

The author acknowledges the use of Google Gemini for language polishing of the manuscript.

## Conflict of interest

The authors declare no conflict of interest.

## Author details

Danilo Malara[1*] and Cristina Pedà[2]

1 Department of Agriculture and Fisheries, Northern Territory Government of Australia, Australia

2 Stazione Zoologica Anton Dohrn, Sicily Marine Center, Messina, Italy

*Address all correspondence to: danilo.malara@szn.it

## References

[1] Coll M, Piroddi C, Steenbeek J, Kaschner K, Ben Rais Lasram F, Aguzzi J, et al. The biodiversity of the Mediterranean Sea: Estimates, patterns, and threats. PLoS One. 2010;**5**(8):e11842

[2] Ben-Tuvia A. The impact of the Lessepsian (Suez Canal) fish migration on the Eastern Mediterranean ecosystem. In: Moraitou-Apostolopoulou M, Kiortsis V, editors. Mediterranean Marine Ecosystems. Boston, MA: Springer US; 1985. pp. 367-375

[3] Millot C. Circulation in the Western Mediterranean Sea. Journal of Marine Systems. 1999;**20**(1):423-442

[4] Bryden HL, Kinder TH. Steady two-layer exchange through the strait of gibraltar. Deep Sea Research Part A Oceanographic Research Papers. 1991;**38**:S445-SS63

[5] Bakun A, Agostini VN. Seasonal patterns of wind-induced upwelling/downwelling in the Mediterranean Sea. Scientia Marina. 2001;**65**(3):243-257

[6] Parras-Berrocal IM, Vazquez R, Cabos W, Sein DV, Álvarez O, Bruno M, et al. Dense water formation in the Eastern Mediterranean under global warming scenario. EGUsphere. 2023;**2023**:1-17

[7] Fry F. Effect of the Environment on Animal Activity. Toronto: University of Toronto Press; 1947

[8] Clarke A, Johnston NM. Scaling of metabolic rate with body mass and temperature in teleost fish. Journal of Animal Ecology. 1999;**68**(5):893-905

[9] Braun CD, Kaplan MB, Horodysky AZ, Llopiz JK. Satellite telemetry reveals physical processes driving billfish behavior. Animal Biotelemetry. 2015;**3**(1):2

[10] Oliver SP, Grothues TM, Mayo ZJ, Williams AL, Silvosa M, Cases G. Depth and temperature profiles reflect individual differences in the daytime diving behaviours of pelagic thresher sharks. Frontiers in Marine Science. 2023;**10**:1128473

[11] Williams CL, Ponganis PJ. Diving physiology of marine mammals and birds: The development of biologging techniques. Philosophical Transactions of the Royal Society B: Biological Sciences. 2021;**376**(1830):20200211

[12] Malara D, Battaglia P, Arostegui MC, Dale JJ, Block BA, Braun CD, et al. Satellite tagging insights into the seasonal movements and behavior of Mediterranean spearfish (Tetrapturus belone, Istiophoridae). Frontiers in Marine Science. 2024;**11**:1362169

[13] Houpert L, Testor P, Durrieu de Madron X, Somot S, D'Ortenzio F, Estournel C, et al. Seasonal cycle of the mixed layer, the seasonal thermocline and the upper-ocean heat storage rate in the Mediterranean Sea derived from observations. Progress in Oceanography. 2015;**132**:333-352

[14] Sabatés A, Olivar MP, Salat J, Palomera I, Alemany F. Physical and biological processes controlling the distribution of fish larvae in the NW Mediterranean. Progress in Oceanography. 2007;**74**(2):355-376

[15] Arostegui MC, Braun CD, Gaube P. Movement and thermal niche of the first satellite-tagged Mediterranean spearfish (*Tetrapturus belone*). Fisheries Oceanography. 2019;**28**(3):327-333

[16] Block BA, Dewar H, Blackwell SB, Williams TD, Prince ED, Farwell CJ, et al. Migratory movements, depth preferences, and thermal biology of Atlantic Bluefin tuna. Science. 2001;**293**(5533):1310-1314

[17] Intergovernmental Panel on Climate C. Climate Change 2022—Impacts, Adaptation and Vulnerability: Working Group II Contribution to the Sixth Assessment Report of the Intergovernmental Panel on Climate Change. Cambridge: Cambridge University Press; 2023

[18] Bianchi CN, Morri C. Global sea warming and "tropicalization" of the Mediterranean Sea: Biogeographic and ecological aspects. Biogeographia–The Journal of Integrative Biogeography. 2003;**24**(1):319-327

[19] Muhling BA, Brill R, Lamkin JT, Roffer MA, Lee S-K, Liu Y, et al. Projections of future habitat use by Atlantic bluefin tuna: Mechanistic vs. correlative distribution models. ICES Journal of Marine Science. 2016;**74**(3):698-716

[20] Erauskin-Extramiana M, Arrizabalaga H, Cabré A, Coelho R, Rosa D, Ibaibarriaga L, et al. Are shifts in species distribution triggered by climate change? A swordfish case study. Deep Sea Research Part II: Topical Studies in Oceanography. 2020;**175**:104666

[21] Laloë J-O, Schofield G, Hays GC. Climate warming and sea turtle sex ratios across the globe. Global Change Biology. 2024;**30**(1):e17004

[22] Hays GC, Laloë J-O, Lee PLM, Schofield G. Evidence of adult male scarcity associated with female-skewed offspring sex ratios in sea turtles. Current Biology. 2023;**33**(1):R14-RR5

[23] Notarbartolo-Di-Sciara G, Zanardelli M, Jahoda M, Panigada S, Airoldi S. The fin whale *Balaenoptera physalus* (L. 1758) in the Mediterranean Sea. Mammal Review. 2003;**33**(2):105-150

[24] Papale E, Pelagatti M, Pedrazzi G, Buscaino G. Occurrence and patterns of fin whale songs reveal alternative migration strategies in Svalbard Islands, Norway. Scientific Reports. 2023;**13**(1):4436

[25] Palmiéri J, Orr JC, Dutay JC, Béranger K, Schneider A, Beuvier J, et al. Simulated anthropogenic $CO_2$ storage and acidification of the Mediterranean Sea. Biogeosciences. 2015;**12**(3):781-802

[26] Vargas CA, Cuevas LA, Broitman BR, San Martin VA, Lagos NA, Gaitán-Espitia JD, et al. Upper environmental $pCO_2$ drives sensitivity to ocean acidification in marine invertebrates. Nature Climate Change. 2022;**12**(2):200-207

[27] Pörtner HO, Peck MA. Climate change effects on fishes and fisheries: Towards a cause-and-effect understanding. Journal of Fish Biology. 2010;**77**(8):1745-1779

[28] Fabry VJ, Seibel BA, Feely RA, Orr JC. Impacts of ocean acidification on marine fauna and ecosystem processes. ICES Journal of Marine Science. 2008;**65**(3):414-432

[29] Lionello P, Malanotte-Rizzoli P, Boscolo R. Mediterranean Climate Variability. 1st ed. Amsterdam, Netherlands: Elsevier; 2006

[30] Cramer W, Guiot J, Fader M, Garrabou J, Gattuso J-P, Iglesias A, et al. Climate change and interconnected risks to sustainable development in the Mediterranean. Nature Climate Change. 2018;**8**(11):972-980

[31] Diffenbaugh NS, Singh D, Mankin JS, Horton DE, Swain DL, Touma D, et al. Quantifying the influence of global warming on unprecedented extreme climate events. Proceedings of the National Academy of Sciences. 2017;**114**(19):4881-4886

[32] Adloff F, Somot S, Sevault F, Jordà G, Aznar R, Déqué M, et al. Mediterranean Sea response to climate change in an ensemble of twenty first century scenarios. Climate Dynamics. 2015;**45**(9):2775-2802

[33] Vargas-Yáñez M, García M, Salat J, Garcia-Martinez M, Pascual J, Moya F. Warming trends and decadal variability in the Western Mediterranean shelf. Global and Planetary Change. 2008;**63**:177-184

[34] MacKenzie BR, Mosegaard H, Rosenberg AA. Impending collapse of bluefin tuna in the Northeast Atlantic and Mediterranean. Conservation Letters. 2009;**2**(1):26-35

[35] Lloret J, Cowx IG, Cabral H, Castro M, Font T, Gonçalves JMS, et al. Small-scale coastal fisheries in European seas are not what they were: Ecological, social and economic changes. Marine Policy. 2018;**98**:176-186

[36] Nowicki RJ, Thomson JA, Fourqurean JW, Wirsing AJ, Heithaus MR. Loss of predation risk from apex predators can exacerbate marine tropicalization caused by extreme climatic events. Journal of Animal Ecology. 2021;**90**(9):2041-2052

[37] Navarro J, Coll M, Somes CJ, Olson RJ. Trophic niche of squids: Insights from isotopic data in marine systems worldwide. Deep Sea Research Part II: Topical Studies in Oceanography. 2013;**95**:93-102

[38] Lejeusne C, Chevaldonne P, Pergent-Martini C, Boudouresque CF, Perez T. Climate change effects on a miniature ocean: The highly diverse, highly impacted Mediterranean Sea. Trends in Ecology & Evolution. 2010;**25**(4):250-260

[39] Laist DW. Impacts of marine debris: Entanglement of marine life in marine debris including a comprehensive list of species with entanglement and ingestion records. In: Coe JM, Rogers DB, editors. Marine Debris: Sources, Impacts, and Solutions. New York, NY: Springer; 1997. pp. 99-139

[40] Kühn S, Bravo Rebolledo EL, van Franeker JA. Deleterious effects of litter on marine life. In: Bergmann M, Gutow L, Klages M, editors. Marine Anthropogenic Litter. Cham: Springer International Publishing; 2015. pp. 75-116

[41] Galgani F, Hanke G, Maes T. Global distribution, composition and abundance of marine litter. In: Bergmann M, Gutow L, Klages M, editors. Marine Anthropogenic Litter. Cham: Springer International Publishing; 2015. pp. 29-56

[42] Pedà C, Battaglia P, D'Alessandro M, Laface F, Malara D, Consoli P, et al. Coupling gastro-intestinal tract analysis with an airborne contamination control method to estimate litter ingestion in demersal elasmobranchs. Frontiers in environmental. Science. 2020;**8**(119):1-13

[43] Pedà C, Romeo T, Panti C, Caliani I, Casini S, Marsili L, et al. Integrated biomarker responses in European seabass *Dicentrarchus labrax* (Linnaeus, 1758) chronically exposed to PVC microplastics. Journal of Hazardous Materials. 2022;**438**:129488

[44] Pedà C, Caccamo L, Fossi MC, Gai F, Andaloro F, Genovese L, et al. Intestinal alterations in European sea bass

*Dicentrarchus labrax* (Linnaeus, 1758) exposed to microplastics: Preliminary results. Environmental Pollution. 2016;**212**:251-256

[45] Nelms SE, Galloway TS, Godley BJ, Jarvis DS, Lindeque PK. Investigating microplastic trophic transfer in marine top predators. Environmental Pollution. 2018;**238**:999-1007

[46] Pietroluongo G, Quintana Martín-Montalvo B, Antichi S, Miliou A, Costa V. First assessment of micro-litter ingested by dolphins, sea turtles and monk seals found stranded along the coasts of Samos Island, Greece. Animals (Basel). 2022;**12**(24):3499

[47] Darmon G, Miaud C, Claro F, Doremus G, Galgani F. Risk assessment reveals high exposure of sea turtles to marine debris in French Mediterranean and metropolitan Atlantic waters. Deep Sea Research Part II: Topical Studies in Oceanography. 2017;**141**:319-328

[48] Nelms SE, Duncan EM, Broderick AC, Galloway TS, Godfrey MH, Hamann M, et al. Plastic and marine turtles: A review and call for research. ICES Journal of Marine Science. 2015;**73**(2):165-181

[49] Romeo T, Battaglia P, Pedà C, Consoli P, Andaloro F, Fossi MC. First evidence of presence of plastic debris in stomach of large pelagic fish in the Mediterranean Sea. Marine Pollution Bulletin. 2015;**95**(1):358-361

[50] Smith M, Love DC, Rochman CM, Neff RA. Microplastics in seafood and the implications for human health. Current Environmental Health Reports. 2018;**5**(3):375-386

[51] Bianchi CN, Morri C. Marine biodiversity of the Mediterranean Sea: Situation, problems and prospects for future research. Marine Pollution Bulletin. 2000;**40**:367-376

[52] FAO. The State of World Fisheries and Aquaculture 2020. Sustainability in Action. Rome, Italy: FAO; 2020

[53] Coll M, Piroddi C, Albouy C, Ben Rais Lasram F, Cheung WWL, Christensen V, et al. The Mediterranean Sea under siege: Spatial overlap between marine biodiversity, cumulative threats and marine reserves. Global Ecology and Biogeography. 2012;**21**(4):465-480

[54] Pauly D, Zeller D. Catch reconstructions reveal that global marine fisheries catches are higher than reported and declining. Nature Communications. 2016;**7**(1):10244

[55] Worm B, Barbier EB, Beaumont N, Duffy JE, Folke C, Halpern BS, et al. Impacts of biodiversity loss on ocean ecosystem services. Science. 2006;**314**(5800):787-790

[56] Ferretti F, Worm B, Britten GL, Heithaus MR, Lotze HK. Patterns and ecosystem consequences of shark declines in the ocean. Ecology Letters. 2010;**13**(8):1055-1071

[57] Sinclair AR, Mduma S, Brashares JS. Patterns of predation in a diverse predator-prey system. Nature. 2003;**425**(6955):288-290

[58] Sinclair AR, Krebs CJ. Complex numerical responses to top-down and bottom-up processes in vertebrate populations. Philosophical Transactions of the Royal Society London B Biological Sciences. 2002;**357**(1425):1221-1231

[59] Ripple WJ, Beschta RL. Trophic cascades in Yellowstone: The first 15 years after wolf reintroduction. Biological Conservation. 2012;**145**(1):205-213

[60] Estes JA, Terborgh J, Brashares JS, Power ME, Berger J, Bond WJ, et al. Trophic downgrading of planet earth. Science. 2011;**333**(6040):301-306

[61] Lotze HK, Worm B. Historical baselines for large marine animals. Trends in Ecology & Evolution. 2009;**24**(5):254-262

[62] Maioli F, Weigel B, Chiarabelli E, Manfredi C, Anibaldi A, Isailović I, et al. Influence of ecological traits on spatio-temporal dynamics of an elasmobranch community in a heavily exploited basin. Scientific Reports. 2023;**13**(1):9596

[63] Dimarchopoulou D, Keramidas I, Sylaios G, Tsikliras AC. Ecotrophic effects of fishing across the Mediterranean Sea. Water. 2021;**13**(4):482

[64] Aoki I, Yamakawa T, Takasuka A. Fish Population Dynamics, Monitoring, and Management: Sustainable Fisheries in the Eternal Ocean. Tokyo: Springer; 2018. 159-76 p

[65] FAO. GFCM 2030 Strategy for Sustainable Fisheries and Aquaculture in the Mediterranean and the Black Sea. Rome: FAO; 2021

Chapter 5

# Impact of Fish Cage Culture (Rainbow Trout, *Oncorhynchus mykiss*) on Zooplankton Structure in Iranian Water of the Caspian Sea

*Siamak Bagheri, Mohammad Sayad Bourani,*
*Azemat Dadai Ghandi and Foong Swee Yeok*

**Abstract**

The Caspian Sea has undergone tremendous changes due mainly to anthropogenic influences. From the 2013 record (January and April), studies were done in the southwestern Caspian Sea (off Jefrud), to gauge the impacts of fish cage culture on the zooplankton community. Scientists found a total of 12 species of zooplankton, belonging to 9 taxa at the study sites. As compared to the 1996–1997 records that marked a total of seven Copepode species at the same study area, only one species was recorded in 2013. The same applied to Cladocera, with only *Pleopis polyphemoides* documented in 2013. *Acartia tonsa* and *Balanus improvisus* were the most dominant species found at most sites, whereas Bivalvia larvae of *Pleopis polyphemoides* and *Synchaeta* sp. were noted only in January of 2013. Zooplankton abundance varied greatly between 1600 and 14,500 ind.m$^{-3}$, dependent on site and period of study. Further check with PCA method confirmed the spatial variations between fish cage culture sites and the references sites. Conclusion is drawn to rectify the impacts of fish cage culture on zooplankton population structure, based on the significant abundance of *A. tonsa*, *B. improvisus*, *P. polyphemoides*, and Bivalvia larvae at the fish cage site as compared to the reference site.

**Keywords:** zooplankton, biodiversity, abundance, decrease, fish, Caspian Sea

## 1. Introduction

The Caspian Sea is a huge internal water body, which expands more than 1200 km measured from north to south, and includes almost 40% of the internal waters on earth [1–3]. Even though it is not connected to any marine system, it is too large to be called a lake [4]. Over the last thirty years, the Caspian Sea has undergone significant ecological alterations [1, 2, 5].

The vertical abundance of zooplankton in the sea is significantly tied to the marine food webs, lending its unparallel roles play in the shallow aquatic ecosystems. The copepods stood out as the most well-noted zooplankton that dominate the Baltic, Black, and Caspian Seas [6, 7]. Zooplankton plays a very important role at the intermediate level of marine food pyramid, as energy transferrer that connects the primary producers (phytoplankton) to the larger consumer such as fish at the higher tropic levels [6, 8].

Anomalies were recorded in the zooplankton community and fish catches in the Caspian Sea since the late 1990s [9], and many research findings pointed to the cause of human activities, climate changes, and the introduction of invasive species such as *Acartia tonsa* Dana, 1849, *Pleopis polyphemoides* Leuckart, 1859, and *Mnemiopsis leidyi* Agassiz, 1865, as the combined culprits [2, 10, 11]. The decrease in zooplankton taxa was shown since 1996 and persisted through 2010 in the southern Caspian Sea [12, 13]. The most notable changes were on the Cladocera, with only two species left in 2010, from the previously recorded nine species [14]. The same reduction in species was also noted in Copepoda, with five species recorded in 1996 and dropped to the sole of *A. tonsa,* as found in 2008 [12–14].

Due to increased demands on seafood, many countries in East and Southeast Asia (China, Japan, India, Malaysia, Thailand, Philippines, and Australia) have shifted to large-scale marine fish cage culture from the traditional wild catch [15]. In Iran, a pioneer fish cage culture was launched in May 2012 by the Iranian Fisheries Organization in the Guilan province (southwestern Caspian Sea), with the rearing of rainbow trout, *Oncorhynchus mykiss,* Walbaum 1792. The fishes were given pelleted food and first harvest was done in October the same year (Unpublished data by Golshhi). Later, Dias et al. [15] and Abery et al. [16] found that fish cage culture practices have raised nutrient levels, caused changes in trophic food web, decreased biodiversity, introduced alien species, and spread diseases in the marine system. In recent years, a number of studies have been conducted on zooplankton structure and variations in the southern Caspian Sea from 1996 to 2010 [9, 11–13, 17, 18]. However, studies to look at the impact of fish cage culture to the intact of Caspian Sea remain scanty. That led to the main objective of this research, which was to assess the enhancements of food/nutrient by the fish cage culture and how that caused the changes in zooplankton structure. The study sites were focused in the southwestern Caspian Sea, off the coast of Jefrud.

## 2. Material and methods

### 2.1 Area studied

Sample collections were done twice in the southwestern Caspian Sea (off the coast of Jefrud) on 7 January and 28 April 2013. No sampling was done between February and March of the same year due to unforeseen circumstances. We selected fish cage sampling stations that were located about 5.5 km away from the shore of Jefrud (N 38°89′ 06″; E 41°51′ 06″). Water samples were collected at three angles from the fish cages: south (S1), northwestern (NW1), and northeastern (NE1). Besides this, a reference station was selected 1 km away from the fish cage site with the same sampling angles: south (S2), northwestern (NW2), and northeastern (NE2) (**Figure 1**). Water samples were drawn at 25 meter below sea surface for both sites. Sampling on all stations were carried out within a day (between 9 am and 2 pm) using a speedboat.

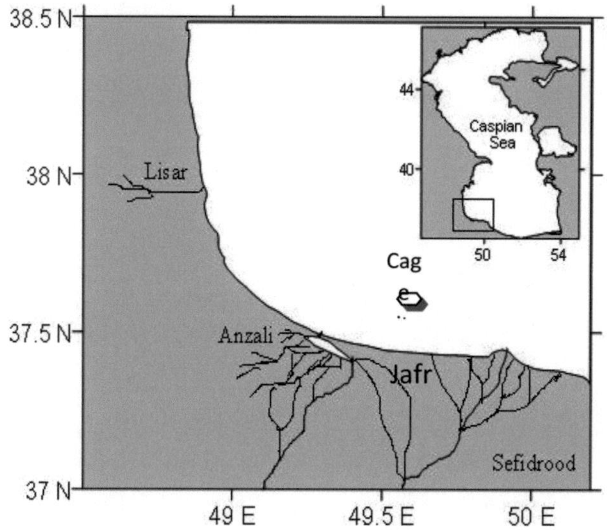


**Figure 1.**
*The sampling stations in the southwestern Caspian Sea, off Jefrud in 2013. Station S1, NW1, & NE1 = near the fish cage site, and S2, NW2, & NE2 = 1 km away from the fish cage site.*

### 2.2 Data collection

We used a Juday net [19] to sample the zooplankton at site, which comes with a 36 cm diameter and 100 $\mu$m mesh size. Vertical hauls were done, from the bottom (25 meter in depth) to the surface of the sea with the Judya net that was hooked onto a handle pulley. Zooplankton samples were preserved in neutral 4% formaldehyde immediately and sent to the laboratory for analysis. About 1 ml Hensen-Stempel pipette was used to collect the subsamples, which were then transferred to a Bogorov chamber for species identification. Following Ref. [20], an inverted microscope was used to count at least 100 individuals for each sample, with life-cycle stages determined. Detailed zooplankton taxonomic classification was based on Ref. [21]. To confirm the differences in zooplankton among the stations, nonparametric analysis (Kruskal-Wallis), followed by the Conover-Inman's test, was used. Further statistical analysis using the principal component analysis (PCA) was also done to determine the dominant zooplankton taxa and its correlation matrix between stations.

## 3. Results

### 3.1 Zooplankton taxa

**Table 1** shows the zooplankton abundance found at the study sites: 12 zooplankton species, belonging to 9 taxa (mero- and holozooplankton). We could not make further discern between mero- and holozooplankton on the fish eggs and larvae since their identification were not determined. Four merozooplankton taxa were recorded: individuals of water spiders (Arachnida), larvae of Bivalvia, Nematoda, and the Cirripedia *Balanus improvisus* as represented by nauplius and cypris larvae (**Table 1**). Another four holozooplankton taxa were: Rotifera (*Synchaeta* sp.), Cladocera (*Pleopis polyphemoides*), and Copepoda (*Ectinosoma concinnum* and *Acartia tonsa*). Copepoda

| | | | Freshwater species | January | | April | | Annual average | | |
|---|---|---|---|---|---|---|---|---|---|---|
| No | | Group | Taxa | average | std | average | Std | total no | average | std |
| 1 | m | Arachnida | Arachnida | 0 | *0* | 0.3 | *0.8* | 2 | 0.166 | *0.57* |
| 2 | m | Bivalvia | Bivalvia larvae | 0 | *0* | 2736.8 | *1889.4* | 16,421 | 1368 | *1915* |
| 3 | m | Cirripedia | *Balanus improvisus* cypris Darwin, 1854 | 53.5 | *22.2* | 53.2 | *31.1* | 640 | 53.33 | *25.74* |
| 4 | m | Cirripedia | *Balanus improvisus* nauplius Darwin, 1854 | 983.3 | *413.8* | 1472.7 | *897.8* | 14,736 | 1228 | *713.8* |
| 5 | h | Cladocera | *Pleopis polyphemoides* Leuckart, 1859 | 0 | *0* | 290.3 | *213.9* | 1742 | 145.2 | *209.3* |
| 6 | h | Copepoda | *Acartia tonsa* nauplius Dana, 1849 | 3068.5 | *576.9* | 968.5 | *826* | 24,222 | 2019 | *1290* |
| 7 | h | Copepoda | *Acartia tonsa* adult Dana, 1849 | 995.6 | *230.7* | 1372.2 | *827.5* | 14,207 | 1184 | *611.6* |
| 8 | h | Copepoda | *Ectinosoma concinnum* Akatova, 1935 | 0 | *0* | 0.2 | *0.4* | 1 | 0.083 | *0.29* |
| 9 | m | Nematoda | Nematoda larvae | 0.5 | *1.2* | 0 | *0* | 3 | 0.25 | *0.87* |
| 10 | | Pisces | Pisces larvae | 0 | *0* | 1.8 | *2.9* | 11 | 0.92 | *2.15* |
| 11 | | Pisces | Pisces ova | 0 | *0* | 1.3 | *2.1* | 8 | 0.66 | *1.55* |
| 12 | h | Rotifera | *Synchaeta* sp. Ehrenberg, 1832 | 0 | *0* | 23.3 | *31.5* | 140 | 11.67 | *24.47* |
| | | | Total (average) | **5102** | *850.2* | **6921** | *4389* | **72,133** | 6011 | *3160* |
| | | | Total (number) | 30,609 | | 41,525 | | 72,133 | | |

**Table 1.**
*Average abundances of zooplankton taxa (ind. $m^{-3}$) according to sampling month, collected in the southwestern Caspian Sea, off Jefrud (2013). h: holoplankton, m: meroplankton.*

made up of 97% of all sampled individuals, followed by Cirripedia and Bivalvia. These were the most abundant groups. In each of these groups, one species was found dominating throughout the two sampling dates: *A. tonsa* and *B. improvisus* (nauplius and cypris larvae), while the Bivalvia larvae dominated the zooplankton community in April (**Table 1**).

### 3.2 Fluctuation of zooplankton

The spatial variations of the zooplankton were recorded as influenced mainly by these three dominant species: *A. tonsa, B. improvisus,* and Bivalvia larvae (**Figure 2**). The highest abundance record of zooplankton (more than 14,500 individuals per $m^3$) was noted in the warmer month of April, at the NE1 (the fish cage site station; **Figure 2a**). The lowest abundance record (1600 $ind.m^{-3}$) was from the NE2 (the reference site station; **Figure 2a**) in the colder month of January. In general,

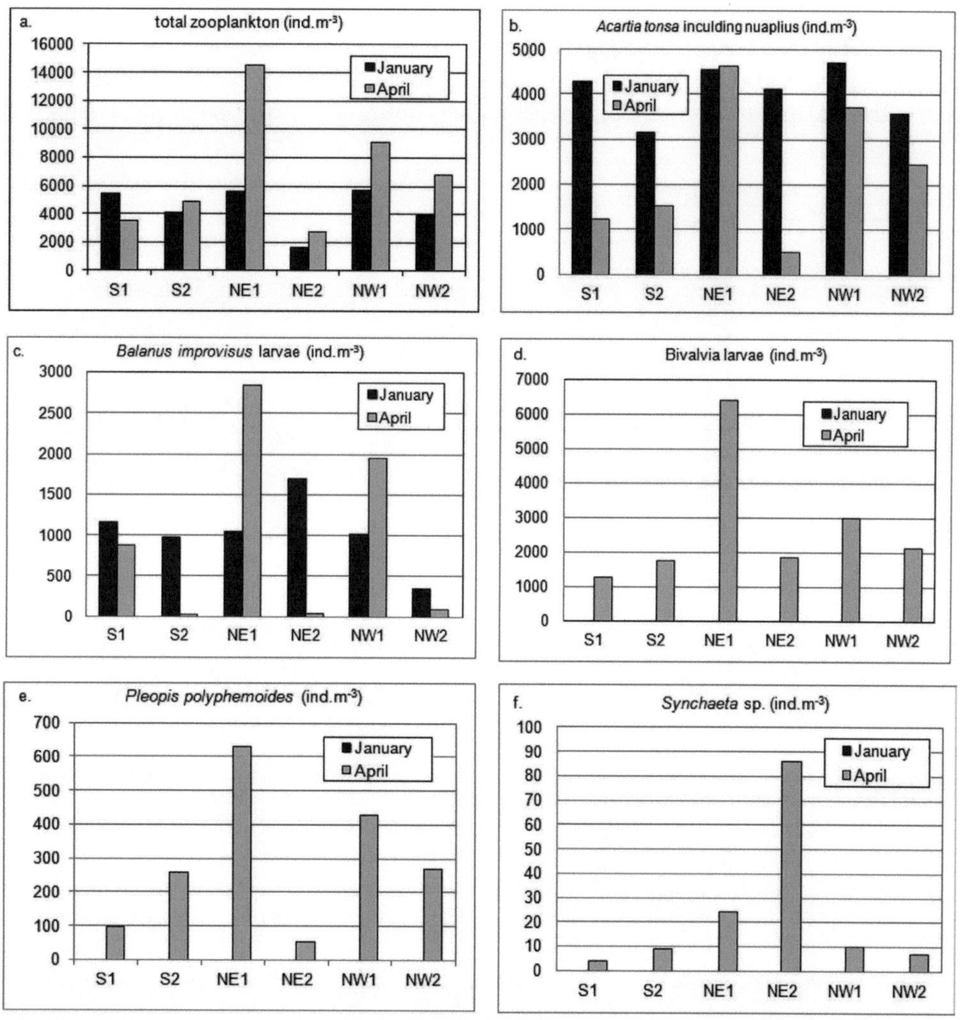


**Figure 2.**
*The sampling stations in the southwestern Caspian Sea, off Jefrud in 2013. S1, NW1, & NE1 = at the fish cage site, and S2, NW2, & NE2 = reference station, 1 km away from the cage site.*

abundance of zooplankton was much higher at the fish cage sites (S1, NW1, & NE1), as compared with the reference site stations (S2, NW2, & NE2). Difference in zooplankton abundances between stations were statistically significant (Kruskal-Wallis; $P < 0.05$).

Warmer month in April played a role in influencing most of the zooplankton abundance at different subsampling location, especially on the *A. tonsa*. The abundance of *A. tonsa* (including nauplius) increased significantly from 3714 to 4620 ind. $m^{-3}$ at the fish cage site stations (NE1 & NW1) in April (**Figure 2b**). While in the colder month of January, there was no significant variation in abundance of *A. tonsa* (700–1200 ind.$m^{-3}$) between the fish cage site stations (S1, NW1, & NE1) and the reference site stations (S2, NW2, & NE2) (**Figure 2b**). However, similar trend was not observed on another species, the Cirripedia. The highest abundance of Cirripedia (*B. improvisus*: nauplius and cypris larvae, 2800 ind.$m^{-3}$) was recorded at the fish cage site stations (NE1) in April, whereas the lowest abundance (only about 30 ind. $m^{-3}$ at the reference site stations (S2 & NE2) was also recorded in the same warmer month in April.

For the Bivalvia larvae (*P. polyphemoides* [Cladocera] and *Synchaeta* sp. [Rotifera]), no sighting was found in the colder month in January (**Figure 2d, e, f**). Recovery was seen in the warmer month of April, with highest abundances of *P. polyphemoides* larvae

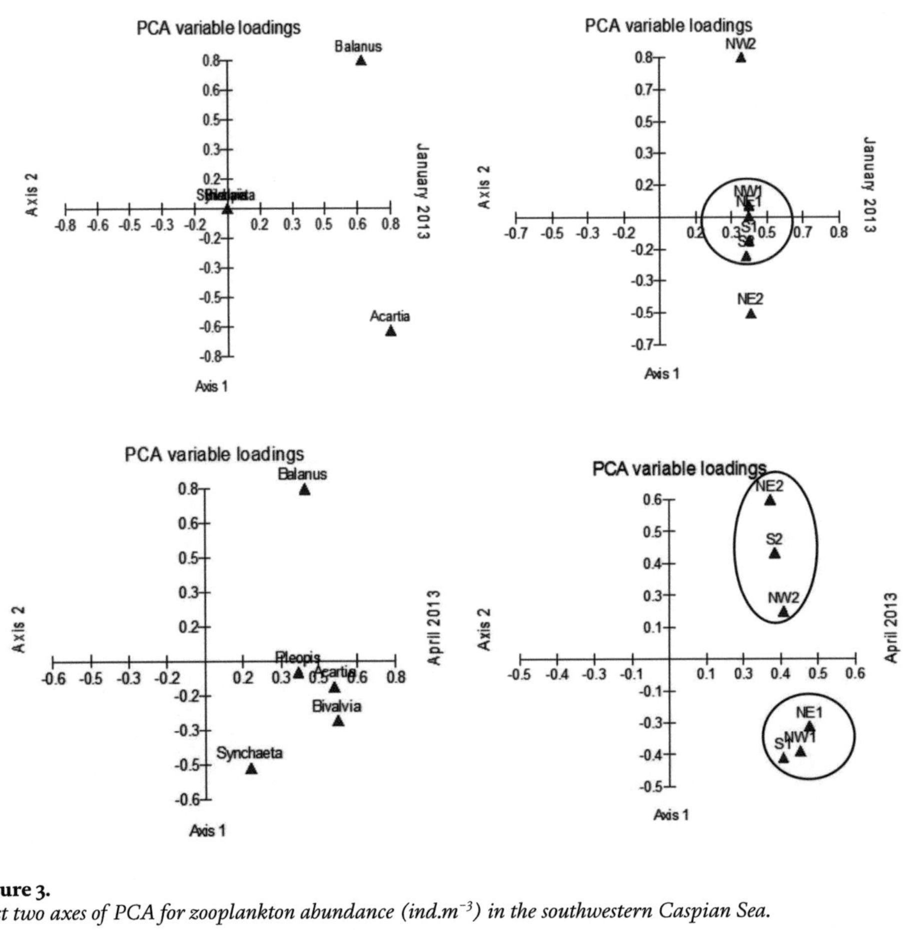


**Figure 3.**
*First two axes of PCA for zooplankton abundance (ind.$m^{-3}$) in the southwestern Caspian Sea.*

amounting to 6500 and 630 ind.m$^{-3}$ at the fish cage site station (NE1) (**Figure 2d, e**). Similarly, *Synchaeta* sp. highest species abundance was recorded (90 ind.m$^{-3}$) at the reference site station in April (NE2; **Figure 2f**).

### 3.3 The PCA analysis based on zooplankton abundances

Further statistical analysis by way of PCA was done on the abundances of these five zooplankton taxa: *A. tonsa*, *B. improvisus*, *Synchaeta* sp., *P. polyphemoides*, and Bivalvia larvae, for the sampling months of January and April 2013. The results show that cumulative values near 99% were achieved for this study. PC1 & PC2 (the two axes where the covariances were grouped based on component loads) as shown in **Figure 3**. Again, *A. tonsa, B. improvisus,* and Bivalvia larvae were found dominating the rest of the taxa with the two highest correlation matrixes.

### 3.4 PCA based on sampling stations

PCA test results on the six-station sampling site (S1, S2, NE1, NE2, NW1, and NW2) show a high of more than 90% cumulative values for zooplankton taxa in this study. The results show that stations near the fish cage sites (S1, NW1, and NE1) were dominated on zooplankton abundances and grouped separately from the sites located in the reference area (S2, NW2, and NE2) (**Figure 3**).

## 4. Discussion

Due to difference in sampling region and dissimilarity of recorded species between research findings, it was tedious to trace the changes in zooplankton abundances with certainty. We take note of the reduction on the number of Cladocera species in 2013 as compared to the older records in southern Caspian Sea [11, 18]. A total of 24 and 13 Cladocera species (including the dominant species *Polyphemus exiguous* Sars, 1897) were known in the southern Caspian Sea in 1996–1997, as compared to just one *P. polyphemoides* that was recorded in 2013 (**Table 1**). All six of the seven recorded Copepoda species (1996–1997) [12] were not found in the 2013 study (**Table 1**) [11]. The 2013 records show that *Pleopis polyphemoides*, *Synchaeta* sp., and Bivalvia larvae were missing in the colder month of January (**Figure 2d, e, f**). These species were noted by Ref. [8] to be usually absent in the summer and winter months (due to extremely low/high temperature), till the year of 2010 with highest abundance occurring in April–May after year 2000 [13, 22]. The findings in this study showed that the abundance of *A. tonsa* including nauplius varied between 2340 and 4064 ind.m$^{-3}$ (**Table 1; Figure 2b**). The average annual abundance of *A. tonsa* presented by others [13, 17, 18] in 1996, 2001–2006, and 2009–2010 revealed a decrease in abundance. The main driving force behind the trend of *A. tonsa* abundance in this study could be related to temporal factor, as the higher abundance that occurred coincided with the warmer April month, as recorded by Ref. [8] in the southwestern Caspian Sea.

It was obvious that fish cage culture has impacted on the zooplankton at the study sites (**Figure 2**), as higher abundances were recorded at the fish cage site stations (S1, NW1, & NE1) as compared with the reference site stations (S2, NW2, & NE2) in this study (**Figure 2a**). As stipulated by Refs. [11, 23] in Caspian Sea and a lake in

Brazil, respectively, zooplankton abundances were strongly affected by high input of nutrients that led to increase in bacteria and small food particles, as might be the case in the 2013 study. Another lake study in Turkey too recorded higher zooplankton abundance near the fish cages with corresponding higher nitrate, phosphate concentrations, and phytoplankton number [15]. Zooplankton and nutrient studies by Refs. [24, 25], respectively, in a lake in China and southern Caspian Sea linked the higher zooplankton abundance (in taxa Cladocera and Copepoda especially) to higher nutrient inputs. Unfortunately, data on nutrient and organic particles were not done in this study. However, PCA drew conclusion on the significance in spatial zooplankton abundance between the high nutrient-load fish cage sampling sites (S1, NW1, & NE1) versus the cleaner reference sites (1 km away from the fish cage site; S2, NW2, & NE2; **Figure 3**).

## 5. Conclusions

We first stipulated that nutrient input from the fish cage culture practice at the study sites might have impacted the abundances of zooplankton in the southwestern Caspian Sea. With the confirmation from all statistical analysis, it is concluded that dominant zooplankton species were significantly more abundant at the stations near the fish cage culture, except station S1 (located at the south of the fish cage, which could be related to the wind direction in different months in 2013). The practice of fish cage culture at the study sites has impacted on the zooplankton population, judging from the striking increase of *B. improvisus*, *P. polyphemoides*, and Bivalvia larvae in abundance at the fish cage culture sites as compared with the reference site.

## Acknowledgements

The authors would like to thank the Inland Waters Aquaculture Institute and the Iranian Fisheries Organization (Guilan province) for financially supporting this project. The assistance received from F. Madadi, E. Yosefzad, and Y. Zahmatkesh in this study is greatly appreciated.

## Author details

Siamak Bagheri[1*], Mohammad Sayad Bourani[1], Azemat Dadai Ghandi[1] and Foong Swee Yeok[2]

1 Inland Waters Aquaculture Research Center, Iranian Fisheries Science Research Institute, Agricultural Research Education and Extension Organization (AREEO), Anzali, Iran

2 School of Biological Sciences, Universiti Sains Malaysia, Penang, Malaysia

*Address all correspondence to: siamakbp@gmail.com

## References


[1] Rodionov SN. Global and Regional Climate Interaction: The Caspian Sea Experience. Dordrecht: Kluwer Academic; 1994. 233 p

[2] Dumont H. Ecocide in the Caspian Sea. Nature. 1995;**377**:673-674. DOI: 10.1038/377673a0

[3] Dumont H. The Caspian Lake: History, biota, structure, and function. Limnology and Oceanography. 1998;**43**:44-52. DOI: 10.4319/lo.1998.43.1.0044

[4] Putans AV, Merklin LR, Levchenko OV. Sediment waves and other forms as evidence of geohazards in Caspian Sea. International Society of Offshore and Polar Engineers. 2010;**20**:241-246

[5] Nasrollahzadeh A. Caspian Sea and its ecological challenges. Caspian Journal of Environmental Sciences. 2010;**8**:97-104

[6] Richardson AJ. In hot water: Zooplankton and climate change. ICES Journal of Marine Science. 2008;**65**:279-295. DOI: 10.1093/icesjms/fsn028

[7] Bagheri S, Mansor M, Wan Maznah WO, Negarestan H. Distribution of *Mnemiopsis leidyi* and zooplankton in the South-Western Caspian Sea, 2008. In: Chung J, editor. Proceedings of the Twentieth International Offshore and Polar Engineering Conference. China: The International Society of Offshore and Polar Engineers (ISOPE) Beijing; 2010. pp. 1014-1018

[8] Bagheri S. Ecological assessment of plankton and effect of alien species in the south-western Caspian Sea. Ecology [thesis]. Penang: Universiti Sains Malaysia; 2012

[9] Daskalov GM, Mamedov EV. Integrated fisheries assessment and possible causes for the collapse of anchovy kilka in the Caspian Sea. ICES Journal of Marine Science. 2007;**64**:503-511. DOI: 10.1093/icesjms/fsl047

[10] Finenko GA, Kideys AE, Anninsky BE, Shiganova TA, Roohi A, Tabari MR, et al. Invasive ctenophore *Mnemiopsis leidyi* in the Caspian Sea: Feeding, respiration, reproduction and predatory impact on the zooplankton community. Marine Ecology Progressive Series. 2006;**314**:171-185. DOI: 10.3354/meps314171

[11] Bagheri S, Niermann U, Mansor M, Foong SY. Biodiversity, distribution and 361 abundance of zooplankton in the Iranian waters of the Caspian Sea off Anzali during 362 1996-2010. Journal of the Marine Biological Association of the UK. 2014;**94**(129-140):363. DOI: 10.1017/S0025315413001288

[12] Roohi A, Yasin Z, Kideys AE, Hwai AT, Khanari AG, Eker-Develi E. Impact of a new invasive ctenophore (*Mnemiopsis leidyi*) on the zooplankton community of the southern Caspian Sea. Marine Ecology. 2008;**29**:421-434. DOI: 10.1007/s10530-009-9648-4

[13] Bagheri S, Niermann U, Sabkara J, Babaei H. State of *Mnemiopsis leidyi* (Ctenophora: Lobata) and mesozooplankton in Iranian waters of the Caspian Sea during 2008 in comparison with previous surveys. Iranian Journal of Fisheries Sciences. 2012;**11**:732-754. DOI: 10.22092/IJFS.2018.114239

[14] Bagheri S, Sabkara J, Mirzajani AR, Khodaparast H, Yosefzad E, Foong SY. List of zooplankton taxa in the Caspian Sea waters of Iran. Journal of Marine

Biology. 2013:1-7. Article ID 134263. DOI: 10.1155/2013/134263

[15] Dias JD, Takahashi EM, Santana NF, Bonecker CC. Impact of fish cage-culture on the community structure of zooplankton in a tropical reservoir Iheringia. Série Zoologia, Porto Alegre. 2011;**101**:75-84. DOI: 10.1590/S0073-47212011000100011

[16] Abery NW, Sukadi F, Budhiman AA, Kartamihardja ES, Koeshendrajana S, Silva SS. Fisheries and cage culture of three reservoirs in West Java, Indonesia: A case study of ambitious development and resulting interactions. Fisheries Management and Ecology. 2005;**12**:315-330. DOI: 10.1111/j.1365-2400.2005.00455.x

[17] Rowshantabari M, Nejatkhah P, Hosseini A, Khodaparast N. Diversity, density and distribution of zooplankton in the southern Caspian Sea in winter 2005 compared with the previous years. Journal of Environmental Science and Technology. 2007;**4**:129-138

[18] Hosseini A. Hydrology and Hydrobiology of the Southern Caspian Sea (1996-1997). Tehran, Iran: Iranian Fisheries Research Organization; 2011. 297 p

[19] Vinogradov ME, Shushkina EA, Musaeva EI, Sorokin PY. A new acclimated species in the Black Sea: The ctenophore *Mnemiopsis leidyi* (Ctenophora: Lobata). Oceanology. 1989;**29**:220-224

[20] Harris RP, Wiebe PH, Lenz J, Skjoldal HR. Zooplankton Methodology Manual Academic Press. UK: Great Britain 2000. 669-684 p. DOI: 10.1016/B978-012327645-2/50014-1

[21] Birshtain YA, Vinogradova LG, Kondakov NN, Koon MS, Astakhova TV, Romanova NN. Invertebrait Atlas Caspian Sea. Moscova: Industry food publisher; 1968

[22] Kasimov AG. Ecology of the Caspian Lake. Baku, Azerbaijan: Elm; 1994

[23] Matsumura TT, Tundisi JG. Plankton richness in eutrophic reservoir (Barra Bonita reservoir, SP, Brazil). Hydrobiologia. 2005;**542**:367-378. DOI: 10.1007/s10750-004-9461-0

[24] Guo L, Li Z. Effects of nitrogen and phosphorus from fish cage-culture on the communities of a shallow lake in middle Yangtze River basin of China. Aquaculture. 2003;**226**:201-212. DOI: 10.1016/S0044-8486(03)00478-2

[25] Bagheri S, Makaremi M. Variation of phytoplankton composition and nutrients near the fish cage-culture in the southern Caspian Sea, Guilan offshore. Journal of Oceanography. 2018;**2018**(9):1-10. DOI: 10.29252/JOC.2018.9.1174

Section 3

# In-Water Biology: Behavior, Ecology and New Technologies

Chapter 6

# Parasites as "Ecological Indicators" for Marine Vertebrates Monitoring and Ecosystem Stability

*Manel Khammassi*


## Abstract

This chapter underscores the importance of integrating parasitological data into marine monitoring programs to enhance conservation and management efforts. By leveraging parasites as ecological indicators, we can improve our understanding of marine ecosystem dynamics and promote the sustainability of marine vertebrate populations.


**Keywords:** parasites, biomarker, marine ecosystem, marine vertebrates, biodiversity

## 1. Introduction

Parasitism is a type of ecological association considered as a long-term relationship between two different organisms where one, the parasite, benefits from the association at the expense of the other, the host, and the parasite harms the host. At the medical level, the costs of the infection-induced are interpreted as diseases. At the ecological level, it is interpreted as reduction in host fitness (longevity, fertility, or both) [1–3]. Parasite-host relationship results from a long-term coevolutionary process characterized by a dynamic feedback loop of adaptations and counter-adaptations [4, 5]. Evolutionary changes in both parasites and hosts are driven by the balance of evolutionary costs and benefits associated with attack and defense strategies. These evolutionary strategies are shaped by the genetic attributes of both the parasite and the host, with each exerting selective pressures on the other, thereby influencing their coevolution [5–9].

In fact, parasitism is a major consumer strategy, and arguably represents the most widespread life-history strategy in nature [5, 10, 11]. By influencing the fitness, and even the behavior of hosts, parasites can regulate host population sizes and affect host communities' structure, and play a highly significant role in regulating host population dynamics [12, 13].

In marine ecosystems, parasites are an essential component of aquatic ecosystems [2, 12, 14]. Their importance in ecosystem functioning is increasingly recognized, and the key role of parasites in the functioning of marine food webs has already been highlighted [15, 16]. In fact, parasites play a crucial role due to their dependence on food webs and complex life cycle interactions that involve a variety of invertebrate and vertebrate hosts. The transmission of parasites through the food chain reflects the diversity of marine organisms and the structure and function of ecosystems [17–19].

Studying these interactions can provide valuable insights into the health of marine environments, making parasites effective indicators for monitoring marine vertebrates. The main goal of this chapter is to highlight the importance of parasites as a novel way to monitor marine vertebrates and ecosystem health.

## 2. Life cycle of parasites in marine ecosystems

Metazoan parasites in marine ecosystems exhibit diverse life cycle strategies to ensure reproduction and survival, broadly categorized into monoxenous (direct) and heteroxenous (indirect) life cycles. In monoxenous cycles, parasites like monogeneans complete their development within a single host species, while heteroxenous cycles involve multiple hosts, such as in the life cycles of digeneans, cestodes, nematodes, and acanthocephalans. These cycles typically start with an invertebrate intermediate host, progress through an intermediate or paratenic host (e.g., fish, squids), and culminate in a definitive host like cetaceans or seabirds, where the parasite reaches maturity and reproduces [20–24]. For instance, the trematode *Haplorchis pumilio* infects various fish species, including threatened ones, such as the IUCN endangered *Etheostoma fonticola*, *Dionda diaboli*, *Gambusia nobilis, Cyprinodon elegans*, and the IUCN vulnerable *Etheostoma grahami*, highlighting its broad host range and ecological significance [25]. Marine parasites generally act as generalists, transferring trophically through predation, though some may kill their hosts to facilitate transmission to the next host [6, 26, 27]. Consequently, parasites move up the food web from lower to higher trophic levels, reflecting predator-prey interactions [13]. Understanding these life cycles and transmission mechanisms is crucial for assessing the ecological roles of parasites and their potential as indicators of environmental health.

### 2.1 Anisakidae life cycle: As model for marine parasites

One of the most completely elucidated life cycles among all marine parasites is the life cycle of nematodes in the Anisakidae family [28]. Nematodes of the genus *Anisakis* exhibit a heterogeneous life cycle, meaning they require multiple hosts to complete their development. Adult nematodes reside in the stomachs of marine mammals (definitive hosts). Eggs are released in feces and hatch into free-swimming larvae in seawater. These larvae are ingested by crustaceans (first intermediate hosts), and they develop further. Fish and squid (second intermediate/paratenic hosts) become infected by consuming these crustaceans. Finally, marine mammals become infected again by consuming fish or squid containing mature larvae. This complex cycle has been extensively studied, with each stage well-characterized [28, 29].

Moreover, the coevolution of parasites and their hosts can provide insights into the evolutionary pressures and adaptations within ecosystems. Studies on *Anisakis* nematodes have shown that coevolutionary dynamics can be used to analyze host population structures and genetic diversity, offering powerful models for ecological and evolutionary research [5, 30–34].

## 3. Role of marine parasites in food webs

The role of parasites in food webs has been ignored for centuries [35]. Recently, several ecological studies have recognized parasites as an integral component of marine ecosystems, intricately woven into the dynamics of food webs through their complex life cycles and interactions with multiple host species at different trophic levels, making them sensitive indicators of changes in food web structure and function [2, 13, 18, 20]. Despite their often-hidden nature and seemingly negligible biomass, parasites exert profound effects on their hosts and, consequently, on the entire food web [3, 36–39].

### 3.1 Transmission routes and life cycles

The complexity and diversity of parasite life cycles present significant challenges. For instance, fewer than 5% of marine helminth life cycles are fully understood, emphasizing the need for further research [40, 41]. Understanding the transmission routes and life cycles of marine parasites is vital for comprehending their impact on food webs. Many marine parasites employ trophic transmission, where an infected host is consumed by the next host in the parasite's life cycle. This mode of transmission is pivotal for unveiling food web links, as it requires a direct feeding interaction between predator and prey. When a vertebrate definitive host, such as a fish, consumes an invertebrate intermediate host harboring a parasite's larval stage, the parasite matures and reproduces within the vertebrate [3]. This trophic linkage is not only critical for the parasite's life cycle but also serves as a natural marker of predation events. Using parasites as tracers may offer several benefits for food web analysis:

a. Direct evidence of predation events: Parasites provide unequivocal evidence of predator-prey interactions. Even for elusive species that are difficult to observe directly, the presence of parasites in their bodies can confirm feeding relationships. For example, discovering a specific parasite in the gut of a predator indicates that the predator consumed a host carrying that parasite [42].

b. Identifying trophic links across spatial scales: Marine parasites can also reveal connections between geographically separated populations. When the same parasite species is found in hosts from different regions, it suggests a link between these populations through migratory or wide-ranging predators. This information is invaluable for understanding the movement and interaction patterns of marine organisms [43].

c. Insights into food web stability and resilience: By analyzing parasite-host relationships, researchers can infer the stability and resilience of marine food webs. Parasites depend on specific hosts and environmental conditions to complete their life cycles. Changes in host populations or environmental disturbances can disrupt these cycles, offering early warning signs of ecosystem stress [44].

d. Enhanced understanding of biodiversity and ecosystem health: The diversity and specificity of parasites reflect the overall health and biodiversity of marine ecosystems. High parasite diversity often correlates with a healthy and diverse host community. Conversely, a decline in parasite diversity may indicate deteriorating environmental conditions or reduced host populations [45].

### 3.2 Impact on hosts

Parasites affect individual hosts through various mechanisms, including mechanical damage, nutrient withdrawal, transfer of microorganisms, toxic effects, and modulation of immune responses. For example, acanthocephalans can perforate fish intestines, cestodes can deplete nutrients, and leeches can transmit blood parasites. These effects can weaken hosts, making them more susceptible to predation and other environmental stresses, thus influencing population dynamics, food web structure, and overall ecosystem stability [12, 46, 47]. In addition, parasites can regulate host populations by affecting host survival, reproduction, and behavior [48].

Interestingly, not all parasitic interactions are harmful. Some parasites provide benefits to their hosts. Such as protozoan and helminth parasites can supply essential nutrients lacking in the host's diet, and ascaridoid nematodes may help in the digestion of large food particles, as they are present in the host's intestine, they can bioaccumulate heavy metal and toxic components [49–53]. As a result, the highest levels of heavy metal accumulation have been recorded in the parasites compared to the host muscle tissues [54, 55]. For example, in nematodes, the Anisakis exhibited the highest accumulation capacity of toxic metals (Cd and Pb) compared to host tissues [56]. Studies on marine mammals in the Neotropics, such as the study on Guiana dolphins (*Sotalia guianensis*) in Brazil, have shown mineralization of the bronchiolar cartilage and an inflammatory process in lungs infected by the nematode *Halocercus brasiliensis*. The presence of this lungworm in dolphins may lead to pulmonary pathologies and death in these cetaceans and may contribute to the decline in the population of these cetaceans [57, 58].

### 3.3 Food web stability

Parasites disproportionately dominate food web links, significantly affecting food web stability. Most traditional food web studies have excluded parasites, leading to potential biases. Empirical data skewed toward free-living species and neglecting parasites have resulted in an incomplete understanding of food web dynamics. Including parasites in food web, models have revealed their substantial impact on food web topology and connectance, providing a more accurate depiction of trophic interactions [18, 35, 41, 59].

Historical and contemporary studies have documented significant effects of parasites on marine host populations [36, 60]. Mass mortalities of fish due to parasite infections have been reported in several studies, such as the destruction of the liver of blue whiting (*Micromesistius potassou*) by larvae of *Anisakis* is a common infection causing deep damage to this organ; as a consequence, the probability of mortality is important [61, 62]. Such events, although rare, demonstrate the potential for parasites to cause large-scale impacts on marine ecosystems. However, under natural conditions, parasite loads are usually below the threshold that would cause widespread host mortality, suggesting that marine ecosystems are generally resilient to parasitic infections [13].

## 4. Marine parasites as ecological indicators

Ecological indicators have been defined as measurable characteristics of the structure (e.g., genetic, population, habitat, and landscape pattern), composition (e.g., genes, species, populations, communities, and landscape types), or function (e.g., genetic, demographic/life history, ecosystem, and landscape disturbance processes) of ecological systems [63]. They have been applied within different ecological levels of organization, from the level of the gene to the landscape [63–65].

Over the past two decades, marine parasites have been increasingly recognized as valuable ecological indicators as they can reflect the impacts of environmental stressors such as pollution, habitat destruction, host population dynamics, and climate change on ecosystems. They can be used to monitor ecosystem processes, biodiversity, and the overall health of habitats [13, 66–68]. Firstly, marine parasites are highly sensitive to environmental changes. For instance, shifts in water temperature, salinity, and pollution levels can significantly affect their life cycles and survival rates [69]. This sensitivity makes them excellent indicators of environmental changes and pollution levels. Secondly, they have complex life cycles involving multiple hosts and stages, which are often intricately linked to specific environmental conditions. Changes in parasite population dynamics can indicate alterations in host populations and environmental conditions [70]. Finally, recent studies have proved that marine parasites can bioaccumulate pollutants from their host organisms, often at higher concentrations than the host tissue. This bioaccumulation makes parasites useful for detecting and measuring environmental contaminants [50]. Based on these features, recent studies classified marine parasites on several categories of indicators [71].

### 4.1 Bioindicators

Bioindicators are organisms used to assess the health of an environment or ecosystem. Marine parasites, due to their sensitivity to changes in environmental conditions, are effective bioindicators. For example, the harbor porpoise (*Phocoena phocoena*), a critically endangered species in the Baltic Sea, has been extensively studied for its parasitic infections. Dzido et al. [72] documented that high parasite loads in harbor porpoises were associated with poor health conditions and increased mortality rates. These findings indicate significant environmental stressors and contribute to understanding the decline of this species.

### 4.2 Biomarkers

Biomarkers are biological responses to environmental chemicals that can be measured in tissues, cells, or fluids, indicating exposure to and effects of pollutants. For example, Kleinertz et al. [73] explored the potential of acanthocephalan fish parasites (*Rhadinorhynchus zhukovi*) recorded in two different host fish species (*Auxis rochei and Auxis thazard*) from Balinese waters as biomarkers using molecular chemical screening techniques. Analysis of the chemical composition of these parasites using pyrolysis-field ionization mass spectrometry (Py-FIMS) revealed distinct molecular signatures, such as volatile alkanes of chain length C15-C17, which are major components of diesel and diesel exhaust particulates. This innovative approach further establishes the role of acanthocephalan parasites as sensitive indicators of ecological change by providing detailed insights into the chemical composition of their host environment.

### 4.3 Bioaccumulators

Bioaccumulators are organisms that accumulate contaminants in their tissues at higher concentrations than those found in their environment or in their prey.

For example, nematode parasites, such as *Hysterothylacium* spp. and *Anisakis* spp., are effective bioaccumulators of heavy metals (e.g., cadmium and lead). Research has shown that these parasites can accumulate higher concentrations of pollutants than their hosts, making them useful for detecting and measuring environmental contaminants [50, 56, 74]. A recent study conducted by [52] in the Gulf of Oman examined the bioconcentration capacities of an endangered ray species, *Himantura cf. Gerrardi*, and critically endangered ray species, *Glaucostegus granulatus*, and their cestod parasites (*Tetragonocephalum* sp., *Polypocephalus* sp., *Rhinebothrium* sp1., and *Rhinebothrium* sp2.) as well as free-living animals including the barnacle and bivalves. Analysis of metal concentrations using graphite furnace atomic absorption spectrometry (GFAAS) revealed that cestodes, especially *Polypocephalus* sp., showed higher levels of contaminants concentrations (cadmium and lead) than in their hosts and other free-living organisms. This high level of bioaccumulation highlights the potential of parasites as indicators of marine pollution.

## 5. Applications in marine vertebrate monitoring

Marine parasites, particularly those with complex life cycles, are increasingly recognized as valuable tools for monitoring the health and stability of marine vertebrate populations, including fish, sharks, mammals, turtles, and seabirds. The completion of their life cycles depends on a stable trophic web, which relies on stable host population sizes.

### 5.1 Health indicators

The presence and intensity of parasitic infections can provide crucial insights into the health and stress levels of marine vertebrates. High parasite loads often indicate weakened host immunity and increased susceptibility to diseases and environmental stressors [75, 76]. For example, a study on the IUCN-endangered sea otters (*Enhydra lutris*) along the California coast has been affected by toxoplasmosis, caused by *Toxoplasma gondii*. This infection has been linked to high mortality rates and impacts on population health and dynamics. The prevalence of toxoplasmosis in sea otters is used as an indicator of land-sea pollution, illustrating the impact of environmental contaminants on marine wildlife [77].

### 5.2 Population monitoring

Parasite data can be instrumental in monitoring population dynamics and migration patterns of marine vertebrates. Specific parasites can reveal migratory routes and feeding grounds of marine mammals and seabirds [27, 44, 78]. For example, the study of the composition and abundance of helminth communities in the IUCN vulnerable loggerhead sea turtles (*Caretta caretta*) in the Mediterranean has provided data reflecting the ontogenetic habitat shift associated with change in their diet [79].

For example, a surveys on the parasitofauna of seabirds underlined that parasite-host dynamics could be altered by marine pollutants. The ingestion of petroleum by seabirds may reduce some gastroenteric parasites, such as digenetics and acanthocephalans, due to its potential anthelmintic effects, and controversly, ciliates and monogenoids might increase. These studies highlights the importance of helminth parasites in monitoring the effects of anthropogenic activities on marine ecosystems [80, 81].

Additionally, parasites, especially nematodes, have been used as biological tags for biomonitoring of the social structure and stock identity of host species, such as fishes (e.g., sardine, herring and mackerel) [29]. Balbuena et al. [82] demonstrated that parasitic infections could provide insights into the social structure and stock differentiation among marine mammal populations. This approach aids in understanding the population dynamics and ecological interactions of these species, which are critical for their conservation and management.

### 5.3 Genetic and ecological insights

Parasites such as *Anisakis* nematodes drive significant genetic differentiation and specialization within both parasite and host populations. The genetic diversity observed in *Anisakis* can be used to analyze host population structures, reflecting connectivity and isolation among host populations [28, 29]. For example, Mattiucci et al. [31] demonstrated that genetic differentiation in *Anisakis simplex* corresponds with distinct herring stocks in the North East Atlantic, underscoring the utility of parasites as indicators for understanding host population dynamics.

Furthermore, the phylogeny and biogeography of parasites like *Anophryocephalus* spp. among pinnipeds reveal how historical events shape current distribution patterns [26]. Studies on microphallid parasites highlight how coevolutionary processes inform our understanding of host ecology and migration, demonstrating the broader applications of parasite monitoring in marine vertebrates [83].

### 5.4 Parasitological data in Franciscana dolphin conservation

Parasites are increasingly recognized as integral components of ecosystems, serving as sensitive indicators of environmental changes and host population health. Gagne et al. [84] highlight the potential of parasites as valuable tools in conservation biology, providing insights into host population dynamics, migration patterns, and environmental stressors. In cetaceans, parasitic infections are particularly informative, as they often reflect the health of the marine environment due to the complex life cycles of these organisms, which are closely tied to their surroundings.

For example, the Franciscana dolphin (*Pontoporia blainvillei*), listed as vulnerable by the IUCN and one of the most threatened cetaceans in the Western South Atlantic, has been the subject of extensive parasitological research [85]. Studies have identified several parasites in the gastrointestinal tract of Franciscana dolphins, including Digenean trematodes (*Synthesium pontoporiae*), Acanthocephalans (*Corynosoma cetaceum* and *Bolbosoma turbinella*), and nematodes (*Anisakis typica* and *A. simplex*). Among these, *Synthesium pontoporiae*, *Corynosoma cetaceum*, and *Anisakis* spp. are recommended as biological markers for Franciscana populations [85–89].

*Synthesium pontoporiae* and *Corynosoma cetaceum* have been identified as particularly useful biological tags, showing similar infection patterns with parasitological indices increasing with latitude across the four Franciscana management areas. These parasites are also suggested as indicators for investigating the existence of distinct Franciscana stocks. Additionally, anisakid nematodes are proposed as the most suitable parasites for detecting latitudinal movements of Franciscana dolphins across different regions [85–89].

The diversity and prevalence of macroparasites in the Franciscana dolphin provide critical insights into the environmental conditions these animals encounter. High parasite loads, for instance, might signal poor health within dolphin populations, potentially linked to factors such as pollution, overfishing, or habitat degradation.

Rocha et al. [85] further discuss the application of parasitological data in monitoring the effects of conservation interventions. By tracking changes in parasite communities over time, conservationists can evaluate the impact of measures such as marine protected areas (MPAs) or fishing regulations. This approach offers a non-invasive method to assess the success of conservation strategies, allowing for the measurement of outcomes without directly disturbing the animals.

## 6. Monitoring marine ecosystem and biodiversity

Marine ecosystems are complex and dynamic environments that are increasingly impacted by anthropogenic activities such as pollution, climate change, and habitat destruction. Monitoring ecosystem health is crucial for maintaining biodiversity and ecological balance. Traditional bioindicators have provided valuable data, but recent studies have highlighted marine parasites as particularly sensitive and informative indicators of ecosystem disturbances and degradation. The use of parasites in environmental monitoring offers a novel approach to understanding and mitigating the impacts of human activities on marine environments [41, 71, 79, 90].

### 6.1 Detecting marine ecosystem disturbance and degradation

#### *6.1.1 Monitoring pollution and environmental health*

##### *6.1.1.1 Pollution indicators*

Monitoring the health of marine ecosystems is essential for their conservation and management. Parasites respond to various environmental stressors, including pollution, which can alter their abundance, diversity, and life cycles [2, 91].

Fish parasites have been used to monitor heavy metals such as mercury, lead, and cadmium in marine environments. Acanthocephalans, cestodes, and nematodes are among the parasites known for their ability to accumulate heavy metals, providing a measure of pollution that integrates over time and across various environmental compartments [54, 92–94].

For instance, the health and condition of Pacific herring (*Clupea pallasi*) in Prince William Sound, Alaska, were monitored using parasitic infections. The study revealed significant correlations between parasite loads and pollution levels, indicating the utility of parasites in detecting environmental stressors [95]. Similarly, flounders (*Paralichthys* spp.) in coastal Chile exhibited impaired health linked to parasitic infections, reflecting the impacts of local pollution [96].

Pollution is not limited to heavy metals. Recent studies have shown that microplastics are among the most harmful pollutants, disrupting ecosystems and increasing organism mortality rates. In fact, the proliferation of microplastics in marine environments poses a significant threat to marine life, impacting organisms from plankton to top predators, including their parasites. Recent research has explored the interactions between microplastics and gut parasites, revealing complex relationships that can have profound ecological implications. The interactions between microplastics and gut parasites in marine fish revealed that microplastics ingestion could impair the immune response of fish, making them more susceptible to parasitic infections.

This interaction underscores the complexity of marine pollution and its multifaceted impacts on marine life [97–99].

Moreover, microplastics can disrupt the gut microbiome of marine organisms, leading to gut dysbiosis and affecting the health and functioning of gut parasites. Chronic exposure to microplastics has been shown to alter the composition and abundance of gut parasites, which can in turn influence host health and susceptibility to other environmental stressors [100].

Additionally, parasites can facilitate the accumulation and transfer of microplastics within marine food webs. For example, studies have found that parasites can bioaccumulate higher concentrations of microplastics compared to their hosts, acting as vectors for these pollutants. This not only affects the parasites and their immediate hosts but also has cascading effects on higher trophic levels and overall ecosystem health [101–103].

By understanding the interactions between microplastics and gut parasites, researchers can gain insights into the broader impacts of plastic pollution on marine ecosystems. This knowledge is crucial for developing effective mitigation strategies to reduce the prevalence of microplastics and protect marine biodiversity.

#### *6.1.1.2 Effects of pollutants on parasite-host interactions*

Marine parasites can also provide insights into the broader ecological impacts of pollution. They can alter host health, immune responses, and behavior, serving as indicators of ecological stress. Carravieri et al. [104] explored how environmental contaminants interact with gastrointestinal parasites in marine predators, revealing the complex dynamics between pollutants and parasite-host interactions. This study demonstrated that contaminants could either increase host susceptibility to infections or reduce parasite loads by affecting the parasites' survival and reproduction.

Similarly, Nissa et al. [105] observed parasitic anomalies in snow trout due to anthropogenic stress, illustrating the significant impact of environmental pollutants on host-parasite interactions. These findings highlight the utility of parasites in detecting ecological changes and stressors, providing a deeper understanding of how pollution affects marine life.

Parasites can bioaccumulate contaminants, exhibit changes in community structure, and affect host health in ways that reflect underlying environmental conditions [91]. These bioaccumulation processes make parasites useful bioindicators for monitoring pollution levels and assessing ecosystem health. Consequently, parasites move up the food web from lower to higher trophic levels, reflecting predator-prey interactions and the ecological impacts of contaminants [6, 13, 26, 27].

#### *6.1.1.3 Pollution monitoring and ecosystem health*

The diversity and abundance of parasite communities can reflect the overall health of ecosystems. High parasite diversity often indicates a healthy, biodiverse environment, while a decline in parasite diversity may signal environmental stress or degradation, habitat loss and/or pollution [90, 106].

The use of fish parasites as bioindicators has been proposed as an early warning system for detecting environmental degradation [107]. Fish parasites are effective in reflecting the ecological conditions of their hosts and can be used to monitor the health of marine ecosystems [90, 92]. For example, in the German Bight, Schmidt et al. used parasites of flounder (*Platichthys flesus*) to monitor ecosystem health [108, 109].

The study revealed that specific parasites, such as nematodes and trematodes, could serve as reliable indicators of pollution. The presence and abundance of these parasites were closely related to pollutant levels, demonstrating their utility in detecting environmental disturbances. In the Red Sea, the intestinal helminth parasites of the siganid fish (*Siganus rivulatus*) were used as bioindicators for trace metal pollution. The study found significant bioaccumulation of heavy metals in the parasites, highlighting the utility of parasites in monitoring environmental contamination [110]. Similarly, the metazoan parasite communities of the shoal flounder (*Syacium gunteri*) have been utilized to monitor chemical contamination in the southern Gulf of Mexico. Vidal-Martínez et al. [111] highlighted how the diversity and abundance of these parasites could indicate levels of pollutants such as heavy metals and organic contaminants. This research underscores the importance of parasites in providing a more comprehensive understanding of environmental pollution and its effects on marine ecosystems.

In addition, parasites of seabirds, such as the common guillemot (*Uria aalge*), have been used to assess the health of marine ecosystems in the North Sea. Changes in parasite prevalence and diversity have been linked to pollution levels and environmental changes [81, 112]. Kleinertz and Palm [113] investigated parasites in grouper fish (*Epinephelus coioides*) from Indonesian coastal ecosystems. They found that parasite diversity and abundance varied significantly between polluted and unpolluted sites. Specific parasites, like the monogenean *Pseudorhabdosynochus coioidesis*, were sensitive to pollution levels, highlighting their potential as indicators of environmental stress and habitat degradation.

In South Africa, studying other benthic marine vertebrates, such as sharks, parasites have been used to monitor metal contamination in marine ecosystems. A recent study conducted on comparisons of concentrations of metals between the tissues of the sharks, the IUCN vulnerable *Rhinobatos annulatus*, the IUCN least concern *Callorhinchus capensis* and *Rhinobatos blochii*, and their endoparasites *Gyrocotyle plana* and *Proleptus obtusus* showed that the tissues of the tapeworm, *G. Plana*, infecting the spiral intestine of *C. capensis* had the highest concentrations of metals (As, Mn, Pb, Ti, and Zn), 2 to 6 times the concentration of the surrounding host tissues. Shark parasites can reflect the levels of environmental contaminants absorbed by their hosts, and could serve as bioindicators of metal pollution in South African marine ecosystem [114].

#### *6.1.2 Parasites and ecosystem functioning*

Parasites occupy various trophic levels and niches within marine ecosystems, making them excellent indicators of ecosystem functioning. They respond to changes in their environment, including pollution, habitat destruction, and climate change, often more rapidly than their hosts or other free-living organisms [2]. They influence host behavior, population dynamics, and energy flow within food webs [3, 47]. Understanding the trophic functioning of parasites provides new insights into ecosystem analysis and health assessment [115]. The impact of parasites on ecosystem processes highlights their importance in ecological studies. They can drive key processes such as nutrient cycling and energy transfer, making them integral components of marine ecosystems [87, 116]. For example, the diversity of trematode larvae was used to assess the health of local fish populations and their habitats. The results indicated that areas with higher parasite diversity were associated with healthier and more stable ecosystems [25, 117].

*DOI: http://dx.doi.org/10.5772/intechopen.1007153*

Marine parasites, especially those with complex life cycles, have coevolved with their definitive hosts, which are often top predators in the food chain, showing high host-parasite specificity [26, 29]. As studying these parasites, tracking food web stability may reflect marine ecosystem state and function [118]. For example, helminths have evolved sophisticated means to ensure their transmission, relying on a variety of marine mammals' hosts, to reproduce [28, 29, 44, 119]. As a consequence, these host-specific parasites are sensitive to the loss of their host species. The disappearance of these parasites can signal a reduction in host diversity and indicate ecosystem degradation and dysfunction [3, 13, 18, 19, 44]. However, in degraded systems, the opposite pattern was observed [120]. Opportunistic and generalist parasites somehow increase in abundance in response to degradation and take advantage of the weakened immune systems of hosts.

On the other hand, as their biological life cycles are complex, parasites integrated into food webs through a variety of invertebrate and vertebrate hosts, that occupy different trophic levels, which may potentially reflect host's trophic position within the food web as well as the presence in the ecosystem of any other organisms that participate in the various parasite life cycles [19].

#### *6.1.3 Parasites and climate change*

Climate change significantly affects marine ecosystems, influencing various biological and ecological processes. One of the key indicators of these changes is the distribution and prevalence of marine parasites. As ocean temperatures rise and salinity levels fluctuate, the life cycles, abundance, and distribution of parasites are directly impacted [90].

For instance, studies in the Antarctic have shown that warming waters have altered the distribution patterns of parasites in marine mammals and fish. Parasites that were previously confined to specific regions are now found in new areas, indicating shifts in host populations and environmental conditions [121]. These changes can disrupt existing parasitic relationships and create new ecological dynamics, affecting the health and survival of marine species.

Researchers discovered tropical coral-infecting parasites in cold marine ecosystems, a finding that underscores the adaptability and widespread distribution of these parasites. This discovery not only expands our understanding of parasite ecology but also suggests that monitoring these parasites can provide insights into environmental changes across different marine habitats [122].

### 6.2 Monitoring marine biodiversity using parasites

The diversity and composition of parasite communities can reflect the health and biodiversity of marine ecosystems [66]. Changes in parasite diversity can reflect the biodiversity and the richness of the host species and their ecological interactions, making them effective indicators of ecosystem health [41, 123]. Recent studies have highlighted the effectiveness of using marine parasites in monitoring biodiversity. One notable study by Wood et al. [124] examined preserved fish specimens from Puget Sound, United States, revealing significant declines in parasite populations over 140 years, correlating with rising sea temperatures. This decline, particularly among parasites with complex life cycles requiring multiple hosts, indicates disruptions in the ecosystem and suggests that parasites are highly sensitive to environmental changes.

Specifically, helminths and trematodes have been shown to play a crucial role in monitoring marine biodiversity and ecosystem health. For example, monogenean parasites affecting coral reef fish, such as *Chaetodon lunulatus*, have been associated with decreased respiratory efficiency and increased susceptibility to environmental stressors. Monitoring these parasitic infections provides valuable information on the health of coral reef ecosystems and the impacts of environmental changes [45, 125].

## 7. Conclusion

Parasites are valuable tools for monitoring marine ecosystem health and pollution. Their sensitivity to environmental changes and their role in ecosystem functioning make them effective bioindicators. By incorporating parasites into ecological monitoring programs, we can gain a deeper understanding of the impacts of pollution and other stressors on marine ecosystems, ultimately aiding in their conservation and management. Despite the potential of parasites as bioindicators, several challenges need to be addressed. The complexity of parasite-host interactions and the variability in parasite responses to pollutants require comprehensive studies to establish standardized methods for their use in environmental monitoring. Additionally, integrating parasitological data with other environmental indicators can enhance the accuracy of ecosystem health assessments and conservation strategies.

## Acknowledgements

The author would like to thank the publisher and the academic editor for the opportunity to take part in this work.

## Conflict of interest

The authors declare no conflict of interest.

## Author details

Manel Khammassi
Research Unit of Wildlife Ecology, Faculty of Sciences of Tunis, University of Tunis El Manar, Tunis, Tunisia

*Address all correspondence to: manel.khammassi@fst.utm.tn

## References

[1] Dobson A, Lafferty KD, Kuris AM, Hechinger RF, Jetz W. Homage to Linnaeus: How many parasites? How many hosts? Proceedings of the National Academy of Sciences. 2008;**105**(supplement_1):11482-11489. DOI: 10.1073/pnas.0803232105

[2] Hatcher MJ, Dunn AM. Parasites in Ecological Communities: From Interactions to Ecosystems. 1st ed. Cambridge: Cambridge University Press; 2011. p. 445

[3] Lafferty KD. Ecosystem consequences of fish parasites. Journal of Fish Biology. 2008;**73**(9):2083-2093. DOI: 10.1111/j.1095-8649.2008.02059.x

[4] Buckingham LJ, Ashby B. Coevolutionary theory of hosts and parasites. Journal of Evolutionary Biology. 2022;**35**(2):205-224. DOI: 10.1111/jeb.13981

[5] Clayton DH, Bush SE, Johnson KP. Coevolution of Life on Hosts: Integrating Ecology and History. 1st ed. Chicago: University of Chicago Press; 2015. p. 320. DOI: 10.7208/9780226302300

[6] Esch GW, Fernandez JC. A Functional Biology of Parasitism: Ecological and Evolutionary Implications. 1st ed. London: Chapman and Hall; 1993. p. 349p

[7] Gandon S, Agnew P, Michalakis Y. Coevolution between parasite virulence and host life-history traits. The American Naturalist. 2002;**160**(3):374-388. DOI: 10.1086/341525

[8] Kurtz J, Schulenburg H, Reusch TB. Host–parasite coevolution—rapid reciprocal adaptation and its genetic basis. Zoology. 2016;**119**(4):241-243. DOI: 10.1016/j.zool.2016.06.011

[9] Lambrechts L, Fellous S, Koella JC. Coevolutionary interactions between host and parasite genotypes. Trends in Parasitology. 2006;**22**(1):12-16. DOI: 10.1016/j.pt.2005.11.008

[10] De Meeûs T, Renaud F. Parasites within the new phylogeny of eukaryotes. Trends in Parasitology. 2002;**18**(6):247-251. DOI: 10.1016/S1471-4922(02)02269-9

[11] Lafferty KD. Interacting parasites. Science. 2010;**330**(6001):187-188. DOI: 10.1126/science.119691

[12] Hatcher MJ, Dick JTA, Dunn AM. How parasites affect interactions between competitors and predators. Ecology Letters. 2006;**9**(11):1253-1271. DOI: 10.1111/j.1461-0248.2006.00964.x

[13] Lafferty KD. Parasites in marine food webs. Bulletin of Marine Science. 2013;**89**(1):123-134. DOI: 10.5343/bms.2011.1124

[14] Poulin R. Are there general laws in parasite ecology? Journal of Parasitology. 2007;**134**(6):763-776

[15] Mundsen P-A, Lafferty KD, Knudsen R, Primicerio R, Klemetsen A, Kuris AM. Food web topology and parasites in the pelagic zone of a subarctic lake. Journal of Animal Ecology. 2009;**78**(3):563-572

[16] Frainer A, McKie BG, Amundsen P-A, Knudsen R, Lafferty KD. Parasitism and the biodiversity-functioning relationship. Trends in Ecology & Evolution. 2018;**33**(4):260-268. DOI: 10.1016/j.tree.2018.01.011

[17] McLaughlin JP, Morton DN, Lafferty KD. Parasites in marine food

webs. In: Behringer DC, Silliman BR, Lafferty KD, editors. Marine disease ecology. 1st ed. Oxford: Oxford University Press; 2020. pp. 45-60. DOI: 10.1093/oso/9780198821632.003.0002

[18] Lafferty KD, Dobson AP, Kuris AM. Parasites dominate food web links. PNAS. 2006;**103**(30):11211-11216. DOI: 10.1073/pnas.0604755103

[19] Thompson RM, Mouritsen KN, Poulin R. Importance of parasites and their life cycle characteristics in determining the structure of a large marine food web. Journal of Animal Ecology. 2005;**74**(1):77-85. DOI: 10.1111/j.1365-2656.2004.00899.x

[20] Bennett J, Presswell B, Poulin R. Tracking life cycles of parasites across a broad taxonomic scale in a marine ecosystem. International Journal for Parasitology. 2023;**53**(5):285-303. DOI: 10.1016/j.ijpara.2023.02.004

[21] Busch MW, Kuhn T, Münster J, Klimpel S. Marine crustaceans as potential hosts and vectors for metazoan parasites. In: Mehlhorn H, editor. Arthropods as Vectors of Emerging Diseases. Parasitology Research Monographs. Vol. 3. Berlin, Heidelberg: Springer; 2012. pp. 329-360. DOI: 10.1007/978-3-642-28842-5_14

[22] Palomba M, Mattiucci S, Crocetta F, Osca D, Santoro M. Insights into the role of deep-sea squids of the genus *Histioteuthis (Histioteuthidae)* in the life cycle of ascaridoid parasites in the Central Mediterranean Sea waters. Scientific Reports. 2021;**11**(1):7135. DOI: 10.1038/s41598-021-86248-5

[23] Huston DC, Ogawa K, Shirakashi S, Nowak BF. Metazoan parasite life cycles: Significance for fish mariculture. Trends in Parasitology. 2020;**36**(12):1002-1012. DOI: 10.1016/j.pt.2020.07.011

[24] Khammassi M, Bahri S, Pekmezci GZ. Morphological and molecular identification of Hysterothylacium larvae (Nematoda: Raphidascarididae) in marine fish from Tunisian Mediterranean coasts. Parasitology Research. 2020;**119**(10):3285-3296. DOI: 10.1007/s00436-020-06848-4

[25] Huston DC, Worsham MD, Huffman DG, Ostrand KG. Infection of fishes, including threatened and endangered species by the trematode parasite *Haplorchis pumilio* (Looss, 1896) (Trematoda: Heterophyidae). BioInvasions Records. 2014;**3**(3):189-194. DOI: 10.3391/bir.2014.3.3.09

[26] Hoberg EP, Am A. Phylogeny, historical biogeography, and ecology of Anophryocephalus spp. (Eucestoda: Tetrabothriidae) among pinnipeds of the Holarctic during the late Tertiary and Pleistocene. Canadian Journal of Zoology. 1992;**70**(4):703-719. DOI: 10.1139/z92-105

[27] Hoberg EP, Brooks DR. A macroevolutionary mosaic: Episodic host-switching, geographical colonization and diversification in complex host–parasite systems. Journal of Biogeography. 2008;**35**(9):1533-1550. DOI: 10.1111/j.1365-2699.2008.01951.x

[28] Mattiucci S, Cipriani P, Paoletti M, Levsen A, Nascetti G. Reviewing biodiversity and epidemiological aspects of anisakid nematodes from the North-east Atlantic Ocean. Journal of Helminthology. 2017;**91**(4):422-439. DOI: 10.1017/S0022149X1700027X

[29] Mattiucci S, Nascetti G. Advances and trends in the molecular systematics of anisakid nematodes, with implications for their evolutionary ecology and host—parasite co-evolutionary processes. Advances in Parasitology.

2008;**66**:47-148. DOI: 10.1016/S0065-308X(08)00202-9

[30] Caballero-Huertas M, Palomba M, Frigola-Tepe X, Muñoz M, Mattiucci S, Viñas J. Ascaridoid parasites in European sardine throughout the annual cycle: Variability in parasitic load according to host stock features. International Journal for Parasitology: Parasites and Wildlife. 2023;**20**:1-11. DOI: 10.1016/j.ijppaw.2022.12.001

[31] Mattiucci S, Giulietti L, Paoletti M, Cipriani P, Gay M, Levsen A, et al. Population genetic structure of the parasite *Anisakis simplex* (s. s.) collected in *Clupea harengus* L. from North East Atlantic fishing grounds. Fisheries Research. 2018;**202**:103-111. DOI: 10.1016/j.fishres.2017.08.002

[32] Anderson RM, May RM. Coevolution of hosts and parasites. Parasitology. 1982;**85**(2):411-426. DOI: 10.1017/S0031182000055360

[33] May RM, Anderson RM. Parasite—host coevolution. Parasitology. 1990;**100**(S1):S89-S101. DOI: 10.1017/S0031182000073042

[34] Papkou A. The influence of population size on host-parasite coevolution [thesis]. Zoological Institute, Kiel University; 2014

[35] Timi JT, Poulin R. Why ignoring parasites in fish ecology is a mistake. International Journal for Parasitology. 2020;**50**(10-11):755-761. DOI: 10.1016/j.ijpara.2020.04.007

[36] Dunne JA, Lafferty KD, Dobson AP, Hechinger RF, Kuris AM, Martinez ND, et al. Parasites affect food web structure primarily through increased diversity and complexity. PLoS Biology. 2013;**11**(6):e1001579. DOI: 10.1371/journal.pbio.1001579

[37] Hatcher MJ, Dick JT, Dunn AM. Parasites that change predator or prey behaviour can have keystone effects on community composition. Biology Letters. 2014;**10**(1):20130879. DOI: 10.1098/rsbl.2013.0879

[38] Marcogliese J. Pursuing parasites up the food chain: Implications of food web structure and function on parasite communities in aquatic systems. Acta Parasitologica. 2001;**46**(2):82-93

[39] McQuaid CF, Britton NF. Parasite species richness and its effect on persistence in food webs. Journal of Theoretical Biology. 2015;**364**:377-382. DOI: 10.1016/j.jtbi.2014.09.035

[40] Blasco-Costa I, Poulin R. Parasite life-cycle studies: a plea to resurrect an old parasitological tradition. Journal of Helminthology. 2017;**91**(6):647-656. DOI: 10.1017/S0022149X16000924

[41] Poulin R, Blasco-Costa I, Randhawa HS. Integrating parasitology and marine ecology: Seven challenges towards greater synergy. Journal of Sea Research. 2016;**113**:3-10. DOI: 10.1016/j.seares.2014.10.019

[42] Thieltges DW, Bordalo MD, Caballero Hernández A, Prinz K, Jensen KT. Ambient fauna impairs parasite transmission in a marine parasite-host system. Parasitology. 2008;**135**(9):1111-1116. DOI: 10.1017/S0031182008004526

[43] Hechinger RF, Lafferty KD. Host diversity begets parasite diversity: Bird final hosts and trematodes in snail intermediate hosts. Proceedings of the Royal Society B: Biological Sciences. 2005;**272**:1059-1066. DOI: 10.1098/rspb.2005.3070

[44] Marcogliese DJ. Parasites of the superorganism: Are they indicators of

ecosystem health? International Journal for Parasitology. 2005;**35**(7):705-716. DOI: 10.1016/j.ijpara.2005.01.015

[45] Hudson PJ, Dobson AP, Lafferty KD. Is a healthy ecosystem one that is rich in parasites? Trends in Ecology & Evolution. 2006;**21**(7):381-385. DOI: 10.1016/j.tree.2006.04.007

[46] Rohde K. Ecology of marine parasites. Helgoländer Wissenschaftliche Meeresuntersuchungen. 1984;**37**:5-33. DOI: 10.1007/BF01989293

[47] Hatcher MJ, Dick JT, Dunn AM. Diverse effects of parasites in ecosystems: Linking interdependent processes. Frontiers in Ecology and the Environment. 2012;**10**(4):186-194. DOI: 10.1890/110016

[48] Friesen OC, Detwiler JT. Parasite-modified chemical communication: Implications for aquatic community dynamics. Frontiers in Ecology and Evolution. 2021;**9**:634754. DOI: 10.3389/fevo.2021.634754

[49] Hursky O, Pietrock M. Intestinal nematodes affect selenium bioaccumulation, oxidative stress biomarkers, and health parameters in juvenile rainbow trout (*Oncorhynchus mykiss*). Environmental Science and Technology. 2015;**49**(4):2469-2476

[50] Khaleghzadeh-Ahangar H, Malek M, McKenzie K. The parasitic nematodes *Hysterothylacium* sp. type MB larvae as bioindicators of lead and cadmium: A comparative study of parasite and host tissues. Parasitology. 2011;**138**(11):1400-1405. DOI: 10.1017/S0031182011000977

[51] Berland B. Are parasites always harmful? In: Proceedings of the Third European Multicolloquium of Parasitology. Vol. 1981. Cambridge, England; 1980. p. 202

[52] Golestaninasab M, Malek M, Roohi A, Karbassi AR, Amoozadeh E, Rashidinejad R, et al. A survey on bioconcentration capacities of some marine parasitic and free-living organisms in the Gulf of Oman. Ecological Indicators. 2014;**37**(A):99-104. DOI: 10.1016/j.ecolind.2013.09.039

[53] Rohde K. Ecology of marine parasites. St. Lucia, Australia: University of Queensland Press; 1982. p. 245

[54] Sures B. The use of fish parasites as bioindicators of heavy metals in aquatic ecosystems: a review. Aquatic Ecology. 2001;**35**:245-255. DOI: 10.1023/A:1011422310314

[55] Malek M, Haseli M, Mobedi I, Ganjali MR, MacKenzie K. Parasites as heavy metal bioindicators in the shark *Carcharhinus dussumieri* from the Persian Gulf. Parasitology. 2007;**134**(7):1053-1056. DOI: 10.1017/S0031182007002508

[56] Nachev M, Sures B. Environmental parasitology: Parasites as accumulation bioindicators in the marine environment. Journal of Sea Research. 2016;**113**:45-50. DOI: 10.1016/j.seares.2015.06.005

[57] de Moura JF, Hauser-Davis RA, Lemos L, Emin-Lima R, Siciliano S. Guiana dolphins (*Sotalia guianensis*) as marine ecosystem sentinels: Ecotoxicology and emerging diseases. In: Whitacre DM, editor. Reviews of Environmental Contamination and Toxicology Volume 228. 1st ed. Cham: Springer; 2014, 2014. pp. 1-29. DOI: 10.1007/978-3-319-01619-1_1

[58] de Oliveira Carvalho Demarque I, de Oliveira FCR, da Silveira LS, Barbosa LA, Ederli NB. The lungworm, Halocercus brasiliensis (Nematoda: Pseudaliidae), from Guiana Dolphins *Sotalia guianensis* from Brazil with Pathological Findings. Journal of Parasitology.

2020;**106**(2):254-260. DOI: 10.1645/19-77

[59] Lafferty KD, Allesina S, Arim M, Briggs CJ, De Leo G, Dobson AP, et al. Parasites in food webs: The ultimate missing links. Ecology Letters. 2008;**11**:533-546. DOI: 10.1111/j.1461-0248.2008.01174.x

[60] Jephcott TG, Alves-de-Souza C, Gleason FH, van Ogtrop FF, Sime-Ngando T, Karpov SA, et al. Ecological impacts of parasitic chytrids, syndiniales and perkinsids on populations of marine photosynthetic dinoflagellates. Fungal Ecology. 2016;**19**:47-58. DOI: 10.1016/j.funeco.2015.03.007

[61] Karasev AB, Mitenev VK, Zubchenko AV, Bezgachina TV. Some data on infestation of blue whiting by nematode *Anisakis* sp. larvae in the northeast Atlantic. International Council for the Exploration of the Sea. 1981;**20**(11):6

[62] Sindermann CJ. Effects of parasites on fish populations: practical considerations. International Journal for Parasitology. 1987;**17**(2):371-382. DOI: 10.1016/0020-7519(87)90112-3

[63] Niemi GJ, McDonald ME. Application of Ecological Indicators*. Annual Review of Ecology, Evolution, and Systematics. 2004;**35**:89-111. DOI: 10.1146/annurev.ecolsys.35.112202.130132

[64] Lausch A, Herzog F. Applicability of landscape metrics for the monitoring of landscape change: Issues of scale, resolution and interpretability. Ecological Indicators. 2002;**2**(1):3-15. DOI: 10.1016/S1470-160X(02)00053-5

[65] Rublee PA, Kempton JW, Schaefer EF, Allen C, Harris J, Oldach DW, et al. Use of molecular probes to assess geographic distribution of Pfiesteria species. Environmental Health Perspectives. 2001;**109**(5):765-767. DOI: 10.1289/ehp.01109s5765

[66] Bennett J, Presswell B, Poulin R. Biodiversity of marine helminth parasites in New Zealand: What don't we know? New Zealand Journal of Marine and Freshwater Research. 2022;**56**(2):175-190. DOI: 10.1080/00288330.2021.1914689

[67] Colin N, Porte C, Fernandes D, Barata C, PadrósF CM, Monroy M, et al. Ecological relevance of biomarkers in monitoring studies of macro-invertebrates and fish in Mediterranean rivers. Science of The Total Environment. 2016;**540**:307-323. DOI: 10.1016/j.scitotenv.2015.06.099

[68] Lauckner G. Ecological effects of larval trematode infestation on littoral marine invertebrate populations. International Journal for Parasitology. 1987;**17**(2):391-398. DOI: 10.1016/0020-7519(87)90114-7

[69] Williams HH, Mackenzie K. Marine parasites as pollution indicators: An update. Parasitology. 2003;**126**(7):S27-S41. DOI: 10.1017/S0031182003003640

[70] MacKenzie K, Abaunza P. Parasites as biological tags for stock discrimination of marine fish: A guide to procedures and methods. Fisheries Research. 1998;**38**(1):45-56. DOI: 10.1016/S0165-7836(98)00116-7

[71] Tramboo SR, Shahardar RA, Wani ZA, Allaie IM, Bulbul KH. Role of parasites as indicators of environ mental pollution: A review. Chemical Science Review and Letters. 2022;**11**(43):311-317. DOI: 10.37273/ chesci.cs-205301429

[72] Dzido J, Rolbiecki L, Izdebska JN, Rokicki J, Kuczkowski T, Pawliczka I. A

global checklist of the parasites of the harbor porpoise *Phocoena phocoena*, a critically-endangered species, including new findings from the Baltic Sea. International Journal for Parasitology: Parasites and Wildlife. 2021;**15**:290-302. DOI: 10.1016/j.ijppaw.2021.07.002

[73] Kleinertz S, Eckhardt K-U, Theisen S, Palm HW, Leinweber P. Acanthocephalan fish parasites (Rhadinorhynchidae Lühe, 1912) as potential biomarkers: Molecular–chemical screening by pyrolysis-field ionization mass spectrometry. Journal of Sea Research. 2016;**113**:51-57. DOI: 10.1016/j.seares.2015.10.001

[74] Pascual S, Abollo E. Accumulation of heavy metals in the whaleworm Anisakis simplex s.l. (Nematoda: Anisakidae). Journal of the Marine Biological Association of the United Kingdom. 2003;**83**(5):905-906. DOI: 10.1017/S0025315403008038h

[75] Díaz-Delgado J, Fernández A, Sierra E, Sacchini S, Andrada M, Vela AI, et al. Pathologic findings and causes of death of stranded cetaceans in the Canary Islands (2006-2012). PLoS One. 2018;**13**(10):e0204444. DOI: 10.1371/journal.pone.0204444

[76] Muniz-Pereira LC, Vieira FM, Luque JL. Checklist of helminth parasites of threatened vertebrate species from Brazil. Zootaxa. 2009;**2123**(1):1-45. DOI: 10.11646/zootaxa.2123.1.1

[77] Miller MA, Gardner IA, Kreuder C, Paradies DM, Worcester KR, Jessup DA, et al. Coastal freshwater runoff is a risk factor for Toxoplasma gondii infection of southern sea otters (Enhydra lutris nereis). International Journal for Parasitology. 2002;**32**(8):997-1006. DOI: 10.1016/S0020-7519(02)00069-3

[78] Williams HH, MacKenzie K, McCarthy AM. Parasites as biological indicators of the population biology, migrations, diet, and phylogenetics of fish. Reviews in Fish Biology and Fisheries. 1992;**2**(2):144-176. DOI: 10.1007/BF00042882

[79] Santoro M, Badillo FJ, Mattiucci S, Nascetti G, Bentivegna F, Insacco G, et al. Helminth communities of loggerhead turtles (*Caretta caretta*) from Central and Western Mediterranean Sea: The importance of host's ontogeny. Parasitology International. 2010;**59**(3):367-375. DOI: 10.1016/j.parint.2010.04.009

[80] de Matos AMRN, Domit C, Bracarense APFRL. Seabirds: Studies with parasitofauna and potential indicator for environmental anthropogenic impacts. Semina: Ciências Agrárias. 2020;**41**(4):1439-1450. DOI: 10.5433/1679-0359.2020v41n4p1439

[81] Khan JS, Provencher JF, Forbes MR, Mallory ML, Lebarbenchon C, McCoy KD. Parasites of seabirds: A survey of effects and ecological implications. Advances in Marine Biology. 2019;**2019**(82):1-50. DOI: 10.1016/bs.amb.2019.02.001

[82] Balbuena JA, Aznar FJ, Fernández M, Raga JA. Parasites as indicators of social structure and stock identity of marine mammals. In: Blix AS, Walløe L, Ulltang Ø, editors. Developments in Marine Biology. Vol. 4. Amsterdam: Elsevier Science; 1995. pp. 133-139. DOI: 10.1016/S0163-6995(06)80017-X

[83] Galaktionov KV, Blasco-Costa I, Olson PD. Life cycles, molecular phylogeny and historical biogeography of the 'pygmaeus' microphallids (Digenea: Microphallidae): Widespread parasites of marine and coastal birds in the Holarctic.

Parasitology. 2012;**139**(10):1346-1360. DOI: 10.1017/S0031182012000583

[84] Gagne RB, Crooks KR, Craft ME, Chiu ES, Fountain-Jones NM, Malmberg JL, et al. Parasites as conservation tools. Conservation Biology. 2022;**36**(1):e13719. DOI: 10.1111/cobi.13719

[85] Rocha A, Valente AL, Marigo J. Chapter 3 - Macroparasites of franciscana: What can their coexistence say to conservationists? In: César Simões-Lopes P, Cremer MJ, editors. The Franciscana Dolphin: On the Edge of Survival. Academic Press; 2022. pp. 49-72. DOI: 10.1016/B978-0-323-90974-7.00021-5

[86] Alves AKM, Souza EAL, Marigo J, Cremer MJ. Intestinal helminths of Franciscana (Pontoporia blainvillei) and Guiana dolphins (Sotalia guianensis) from the north coast of Santa Catarina state, Brazil. Oecologia Australis. 2017;**21**(1):83-87

[87] Hoss D, Valente AL, Ott PH. Gastrointestinal helminths of the franciscana Pontoporia blainvillei (Mammalia: Cetartiodactyla) in the Northern coast of Rio Grande do Sul State. Southern Brazil. Oecologia Australis. 2017;**21**(1):72-82. DOI: 10.4257/oeco.2017.2101.08

[88] Rocha AA. Comunidade Componente de Helmintos Gastrointestinais de Pontoporia Blainvillei (Gervais & d'Orbigny, 1844) do Litoral sul do Rio Grande do Sul relação com a Dieta e Determinação de Estoques Ecológicos. MSc Thesis. Instituto de Oceanografia da Universidade Federal do Rio Grande; 2010. p. 133

[89] Seabra CB. Helmintos gastrointestinais de toninha, Pontoporia blainvillei (Cetartiodactyla, Pontoporiidae), do litoral de São Paulo. Brasil. MSc Thesis: Universidade Estadual Paulista, Campus do Litoral Paulista, Instituto de Biociências; 2018. p. 60

[90] Palm HW. Fish parasites as biological indicators in a changing world: Can we monitor environmental impact and climate change? In: Mehlhorn H, editor. Progress in Parasitology. Parasitology Research Monographs. Vol. 2. Berlin, Heidelberg: Springer; 2011. pp. 223-250. DOI: 10.1007/978-3-642-21396-0_12

[91] Sures B, Nachev M, Selbach C, Marcogliese DJ. Parasite responses to pollution: what we know and where we go in 'Environmental Parasitology'. Parasites and Vectors. 2017;**10**:1-19. DOI: 10.1186/s13071-017-2001-3

[92] Khan R. Parasites of fish as biomarkers of environmental degradation: A field study. Bulletin of Environmental Contamination and Toxicology. 2004;**72**(2)

[93] Mehana E-SE, Khafaga AF, Elblehi SS, Abd El-Hack ME, Naiel MAE, Bin-Jumah M, et al. Biomonitoring of heavy metal pollution using acanthocephalans parasite in ecosystem: An updated overview. Animals. 2020;**10**(5):811. DOI: 10.3390/ani10050811

[94] Najm M, Fakhar M. Helminthic parasites as heavy metal bioindicators in aquatic ecosystems. Medical Laboratory Journal. 2015;**9**(4):26-32

[95] Elston RA, Drum AS, Pearson WH, Parker K. Health and condition of pacific herring *Clupea pallasi* from Prince William Sound, Alaska, 1994. Diseases of Aquatic Organisms. 1997;**31**(2):109-126. DOI: 10.3354/dao031109

[96] George-Nascimento M, Khan RA, Garcias F, Lobos V, Muñoz G, Valdebenito V. Impaired health

in flounder, *Paralichthys* spp. Inhabiting Coastal Chile. Bulletin of Environmental Contamination and Toxicology. 2000;**64**:184-190. DOI: 10.1007/s001289910028

[97] Parker B, Andreou D, Green ID, Britton JR. Microplastics in freshwater fishes: Occurrence, impacts and future perspectives. Fish and Fisheries. 2021;**22**(3):467-488. DOI: 10.1111/faf.12528

[98] Pennino MG, Bachiller E, Lloret-Lloret E, Albo-Puigserver M, Esteban A, Jadaud A, et al. Ingestion of microplastics and occurrence of parasite association in Mediterranean anchovy and sardine. Marine Pollution Bulletin. 2020;**158**:111399. DOI: 10.1016/j.marpolbul.2020.111399

[99] Hernandez-Milian G, Lusher A, MacGabban S, Rogan E. Microplastics in grey seal (*Halichoerus grypus*) intestines: Are they associated with parasite aggregations? Marine Pollution Bulletin. 2019;**146**:349-354. DOI: 10.1016/j.marpolbul.2019.06.014

[100] Fackelmann G, Sommer S. Microplastics and the gut microbiome: how chronically exposed species may suffer from gut dysbiosis. Marine Pollution Bulletin. 2019;**143**:193-203. DOI: 10.1016/j.marpolbul.2019.04.030

[101] Sheehan KL, Patel P, Mertz Z. Parasites and Plastics in and on Stranded Northern Gannets, *Morus bassanus*. Matters. 2021;**7**(6):e202105000003. DOI: 10.13140/RG.2.2.10489.80484

[102] McHale ME, Sheehan KL. Bioaccumulation, transfer, and impacts of microplastics in aquatic food chains. Journal of Environmental Exposure Assessment. 2024;**3**:15. DOI: 10.20517/jeea.2023.49

[103] Zhang E, Kim M, Rueda L, Rochman C, VanWormer E, Moore J, et al. Association of zoonotic protozoan parasites with microplastics in seawater and implications for human and wildlife health. Scientific Reports. 2022;**12**(1):6532. DOI: 10.1038/s41598-022-10485-5

[104] Carravieri A, Burthe SJ, de La Vega C, Yonehara Y, Daunt F, Newell MA, et al. Interactions between environmental contaminants and gastrointestinal parasites: novel insights from an integrative approach in a marine predator. Environmental Science and Technology. 2020;**54**(14):8938-8948. DOI: 10.1021/acs.est.0c03021?ref=pdf

[105] Nissa NU, Jan M, Tantray JA, Dar NA, Jan A, Ahmad F, et al. Parasitic anomalies observed in snow trout due to anthropogenic stress in water bodies. Saudi Journal of Biological Sciences. 2022;**29**(4):2921-2925. DOI: 10.1016/j.sjbs.2022.01.022

[106] Zargar UR, Chishti MZ, Rather MI, Rehman M. Parasites as emerging biomonitoring tools-promises and pitfalls. Proceedings of the National Academy of Sciences, India Section B: Biological Sciences. 2022;**92**(4):731-739. DOI: 10.1007/s40011-022-01406-7

[107] Mackenzie K. Parasites as pollution indicators in marine ecosystems: a proposed early warning system. Marine Pollution Bulletin. 1999;**38**(11):955-959. DOI: 10.1016/S0025-326X(99)00100-9

[108] Schmidt V, Zander S, Körting W, Broeg K, von Westernhagen H, Dizer H, et al. Parasites of flounder (*Platichthys flesus* L.) from the German Bight, North Sea, and their potential use in biological effects monitoring: C. Pollution effects on the parasite community and a comparison to biomarker responses. Helgoland Marine

Research. 2003;**57**:262-271. DOI: 10.1007/s10152-003-0159-x

[109] Schmidt V, Zander S, Körting W, Steinhagen D. Parasites of the flounder *Platichthys flesus* (L.) from the German Bight, North Sea, and their potential use in ecosystem monitoring: A. Infection characteristics of potential indicator species. Helgoland Marine Research. 2003;**57**:236-251. DOI: 10.1007/s10152-003-0147-1

[110] Al-Hasawi ZM. Environmental Parasitology: intestinal helminth parasites of the siganid fish Siganus rivulatus as bioindicators for trace metal pollution in the Red Sea. Parasite. 2019;**26**:12. DOI: 10.1051/parasite/2019014

[111] Vidal-Martínez VM, Centeno-Chalé OA, Torres-Irineo E, Sánchez-Ávila J, Gold-Bouchot G, Aguirre-Macedo ML. The metazoan parasite communities of the shoal flounder (*Syacium gunteri*) as bioindicators of chemical contamination in the southern Gulf of Mexico. Parasites and Vectors. 2014;7:1-14. DOI: 10.1186/s13071-014-0541-3

[112] Schwantes U. Impact of anthropogenous environmental factors on the marine ecosystem of trophically transmitted helminths and hosting seabirds: Focus on North Atlantic, North Sea, Baltic and the Arctic seas. Helminthologia. 2023;**60**(4):300-326. DOI: 10.2478/helm-2023-0034

[113] Kleinertz S, Palm HW. Parasites of the grouper fish *Epinephelus coioides* (*Serranidae*) as potential environmental indicators in Indonesian coastal ecosystems. Journal of Helminthology. 2015;**89**(1):86-99. DOI: 10.1017/S0022149X1300062X

[114] Morris T, Avenant-Oldewage A, Lamberth S, Reed C. Shark parasites as bio-indicators of metals in two South African embayments. Marine Pollution Bulletin. 2016;**104**(1-2):221-228. DOI: 10.1016/j.marpolbul.2016.01.027

[115] Arias-González JE, Morand S. Trophic functioning with parasites: a new insight for ecosystem analysis. Marine Ecology Progress Series. 2006;**320**:43-53. DOI: 10.3354/meps320043

[116] Sures B, Nachev M, Pahl M, Grabner D, Selbach C. Parasites as drivers of key processes in aquatic ecosystems: Facts and future directions. Experimental Parasitology. 2017;**180**:141-147. DOI: 10.1016/j.exppara.2017.03.011

[117] Stout L, Garenne A, de Montaudouin X. Marine trematode parasites as indicators of environmental changes. Ecological Indicators. 2022;**141**:109089. DOI: 10.1016/j.ecolind.2022.109089

[118] Gibson DI, Harris EA, Bray RA, Jepson PD, Kuiken T, Baker JR, et al. A survey of the helminth parasites of cetaceans stranded on the coast of England and Wales during the period 1990-1994. Journal of Zoology. 1998;**244**(4):563-574. DOI: 10.1111/j.1469-7998.1998.tb00061.x

[119] Pool R, Shiozaki A, Raga JA, Fernández M, Aznar FJ. Molecular phylogeny of the *Pseudaliidae* (*Nematoda*) and the origin of associations between lungworms and marine mammals. International Journal for Parasitology: Parasites and Wildlife. 2023;**20**:192-202. DOI: 10.1016/j.ijppaw.2023.03.002

[120] Dušek L, Gelnar M, Šebelová Š. Biodiversity of parasites in a freshwater environment with respect to pollution: metazoan parasites of chub (*Leuciscus cephalus* L.) as a model for statistical

evaluation. International Journal for Parasitology. 1998;**28**(10):1555-1571. DOI: 10.1016/S0020-7519(98)00130-1

[121] Bestley S, Ropert-Coudert Y, Bengtson NS, Brooks CM, Cotté C, Dewar M, et al. Marine ecosystem assessment for the Southern Ocean: birds and marine mammals in a changing climate. Frontiers in Ecology and Evolution. 2020;**8**:566936. DOI: 10.3389/fevo.2020.566936

[122] Trznadel M, Holt CC, Livingston SJ, Kwong WK, Keeling PJ. Coral-infecting parasites in cold marine ecosystems. Current Biology. 2024;**34**(8):1810-1816. DOI: 10.1016/j.cub.2024.03.026

[123] MacKenzie K, Hemmingsen W. Parasites as biological tags in marine fisheries research: European Atlantic waters. Parasitology. 2015;**142**(1):54-67. DOI: 10.1017/S0031182014000341

[124] Wood CL, Welicky RL, Preisser WC, Leslie KL, Mastick N, Greene C, et al. A reconstruction of parasite burden reveals one century of climate-associated parasite decline. National Academy of Sciences of the United States of America. 2023;**120**(3):e2211903120. DOI: 10.1073/pnas.2211903120

[125] Reverter M, Sasal P, Suzuki MT, Raviglione D, Inguimbert N, Pare A, et al. Insights into the natural defenses of a coral reef fish against gill ectoparasites: Integrated metabolome and microbiome approach. Metabolites. 2020;**10**(6):227. DOI: 10.3390/metabo10060227